A Mathematical Mystery Tour

Discovering the Truth and Beauty of the Cosmos

A. K. Dewdney

John Wiley & Sons, Inc.

New York • Chichester • Weinheim • Brisbane • Toronto • Singapore

Published by John Wiley & Sons, Inc.
Published simultaneously in Canada

Designed and produced by Navta Associates, Inc.

Library of Congress Cataloging-in-Publication Data:
Dewdney, A. K.
A mathematical mystery tour : discovering the truth and beauty of the cosmos / A. K. Dewdney
p. cm.
Includes index.
ISBN 0-471-23847-3 (cloth : alk. paper)
ISBN 0-471-40734-8 (paper)
1. Mathematics. I. Title.
QA36.D485 1999
510-dc21 98-36470

Printed in the United States of America
10 9 8 7 6 5 4 3 2 1

Contents

PART III
THE VANISHING ACT

PART IV
THE ENGINES OF THOUGHT

Preface

Pythagoras, the Greek mathematician and philosopher who flourished in the fifth century B.C., has always ranked among the world's greatest mathematicians, sometimes as the greatest. Two of his most important achievements—the discovery of incommensurate magnitudes and the Pythagorean theorem—play vital roles in the foundations of mathematics.

Despite his fundamental insights into mathematics, Pythagoras had another side that is only partly explained by the culture of his time and place. In particular, he had a belief about the cosmos and its relation to mathematics that seems altogether odd. Today, many scientists believe that mathematics has a striking relationship with reality. A few scientists even believe that mathematics in some sense governs or controls reality. But who could possibly believe that mathematics *makes* reality? Pythagoras did.

Pythagoras was not exactly a scientific prophet. His ponderings in natural philosophy (now called "natural science") have not all survived objective scrutiny. For example, Pythagoras also believed that the Sun went around the Earth and that "below" the Earth (which he understood to be round) was a great "central fire." To that place would the god Apollo repair after sunset to replenish his chariot and horses of fire before rising again into the dawn. We cannot rule out the possibility that Pythagoras understood Apollo as a metaphor, but how was he to know that it was no ordinary fire that burned in the Sun, but an atomic one? As for the Sun going around the Earth, in Pythagoras's day, everyone believed that!

I have often wondered what Pythagoras would make of modern

science and mathematics. He could hardly fail to be pleased at the key role played by the Pythagorean theorem in almost every branch of mathematics and, therefore, science. Nonetheless, would he not gently ask about "therefore"? Would he not inquire into our progress with the central question of his scientific life? "Have you demonstrated yet that the fabric of the cosmos is woven of mathematics?"

As a tribute to the founder of modern mathematics, I have decided to take up this question on Pythagoras's behalf, some 2,500 years after his death. My approach can only be called "novel," being the fictional narrative of a journey to four corners of the world. The mathematical odyssey herein explores two key questions about mathematics and its relationship to reality: Why is mathematics so amazingly successful in describing the structure of physical reality? Is mathematics created, or is it discovered? Such questions circle the central fire of my Pythagorean quest. The answers given by the four characters illuminate the subject from four different directions. They lead to some tentative but startling conclusions that directly confront a modern tendency, even among some scientists, to sweep questions about reality under a post-modern rug.

Pythagoras was a mystic as well as a mathematician. I use the word *mystic* in the technical, traditional sense, not the modern pejorative one. In other words, Pythagoras believed that some truths could be achieved by direct contemplation after a suitable (and very rigorous) preparation of the body and mind. He founded a mystical tradition called the Pythagorean brotherhood. Not only did the brotherhood survive some 1,000 years, well into the Islamic era, but Pythagoreans keep popping up in our narrative, right into the nineteenth century. The more contemporary Pythagoreans were not mystics, as far as I know, merely prominent scientists who described themselves as "Pythagoreans." They believed, at the very least, that physical reality has a mathematical basis.

We do not need to be mystics to believe this much. The whole purpose of this adventure is to breathe new life into ancient questions. It is just possible that the next great scientific paradigm shift will involve them directly. Only then will Pythagoras find rest.

Point of Departure

Paris, June 20, 1995

I sit in the departure lounge for Air France Flight 372, headed for Athens. I arrived here only an hour ago on a transatlantic flight. Apart from a general lassitude, I am in remarkably good shape.

I can see our next plane from the lounge window. With all the sleek beauty demanded by supersonic flight, it is a symbol of technology in the late twentieth century. Such technology, I remind myself, depends almost entirely on science, and science—particularly physical science—depends almost entirely on mathematics. It seems that I will fly to Athens by the sheer power of mathematics. The turbine blades will turn in circles, the rearward force of the jet exhaust will produce an equal and opposite forward thrust, airframe components will resist stress in proportion to their cross sections, and the thin stratospheric air will slip over wings mathematically optimized for lift, exactly equaling gravity.

That, I remind myself, is partly what this trip is all about: the power of mathematics, its amazing applicability in science and technology. The coming flight is but the first step in a long journey.

I have appointments in Turkey, Jordan, Italy, and England to meet various thinkers, some eminent, some unknown. I hope they will throw some light on the question I am pursuing: What is the true nature of mathematics?

The question is hopelessly vague, of course, but I have thought of two more highly focused questions that, if answered, will go a long way to settling the first one.

1. Why is mathematics so incredibly useful in the natural sciences?
2. Is mathematics discovered, or is it created?

I cannot escape the feeling that the two questions are related, perhaps closely related. Precisely how they are related, however, I don't know. To answer these two questions and, with luck, to understand how they are related is the purpose of my quest.

There can be no doubt that most scientists, particularly physical scientists, find mathematics not only useful but also indispensable. It enables us to plot the positions of planets years in advance, to predict the orbitals of electrons, to describe the airflow over the wing of the aircraft I am about to take. The equations and formulas of physics and other inductive sciences, as well as the axioms and theorems on which they are based, all too often describe physical realities with startling precision and universality. All too often, they lead to accurate predictions of new phenomena, whether particles or planets—and all too often, they result in amazing machines that work perfectly. Such power cries out for an explanation.

Unless the amazing power of mathematics to describe physical reality is a case of coincidence or massive self-deception, the first question I have posed almost rephrases itself. *Why* is the physical universe determined (or accurately describable) to so great an extent by mathematical *ideas*? I think it strange that people rarely ask this question, yet it draws me like a lodestar.

As for the second question, whether mathematics is discovered or created, I must confess to an almost mystical experience when I work on a mathematical problem, an experience that seems imposed from without. I often have an eerie feeling that the answer lies *out there*, somehow, that it exists whether I think about it or not. When a solution comes (if it comes), I have the overwhelming impression of having discovered it, rather than having created it. Each of us is entitled to our personal impressions, but how can I defend this view objectively? I could point out that when an answer comes (if it comes) to someone else working on the same problem I have solved, it is all too often the same solution. That person may express the solution differently, but the two solutions will be mutually transformable, equivalent, or, in a word, the same. The solution is out there, somewhere, waiting.

At its most fundamental level, mathematics is about truth. How can truth be created? Much of mathematics has an extraordinary beauty as well, but this does not appear to be a fundamental feature. In mathematics, beauty is not synonymous with truth, although some truths are beautiful and some beautiful things are true. A mathematician may compose a beautiful theorem as a work of art, only to discover that it is not true, which absolutely dooms it as mathematics.

Also, although some theorems have an extraordinary beauty, others are quite ugly. It cannot be asserted that mathematicians choose in any sense what will be true and what not. The creative act of the mathematician, if there is one, is to imagine what might be true, conjecturing the shape of a theorem to be found, then bending every effort to finding it, exactly like an explorer. The quest is not guaranteed to succeed, however. It depends on something besides the mathematician's efforts, something else entirely. It depends on what is true.

If we allow for the moment that mathematics is not created, but discovered, we must ask, *Why?* Does mathematics have an independent existence? These are old chestnuts, mulled over by mathematicians through the ages but largely unresolved and today nearly forgotten. Who, besides the few scholars I have arranged to meet, works on them today?

It is said that mathematicians indulge in their philosophy of choice on holidays, but the rest of the time they are Platonists. For

Plato, mathematics pointed to a world of pure and indestructible forms or archetypes, which enjoy a kind of superreality. In *The Republic,* he invites us to a cave where he has built a fire. We sit with our backs to the fire, facing a wall, like a movie-theater audience. Behind us, Plato superimposes statues and other forms between the fire and the wall, producing recognizable shadowy shapes. As we watch the shadow movie on the wall, we hear Plato paraphrased on the sound track: "As you go about the world, seeing trees and stones and birds, you are really just sitting in my cave, seeing but the shadows of higher realities on the wall of your perception." Every tree, every stone, every bird is but the projection of an archetypal tree, stone, or bird, its form outlined by a kind of Olympian light.

Mathematicians, even those who describe themselves as Platonists, do not go this far. Their only archetypes, as such, are mathematical in nature. Every circle one might draw is but the manifestation of a perfect, ideal circle. It has no thickness and is therefore invisible. It has no particular location and is therefore unlocatable. If we draw circles that are increasingly accurate and refined, with larger radii and thinner lines, we come closer to the ideal circle but never quite realize it. We think not about our drawn circle but about the ideal archetype, in effect.

We know a lot about this ideal circle. We know, for example, that the ratio of its circumference to its diameter is the transcendental number pi, which has an infinite number of digits.

$$\text{pi} = 3.14159265358979\ldots$$

If we measure the circles we draw, their circumferences and diameters, taking ratios as we go, we come closer and closer to the real value of pi, 3.14, then 3.142, then 3.1416, and so on.

All mathematical concepts share this fundamental property. Not only every circle, but also every line, every written number, every symbolic expression points to an ideal concept, an abstraction that can only be grasped through such drawings and symbols. Most mathematicians do not suppose that mathematics is the very source of reality. Nonetheless, many of them have the same impression, as I do, that mathematics has an independent existence of sorts, amounting almost to a place that one can explore.

Whatever might be said about the independent existence of Plato's world, this much is certain: The truths of mathematics do not obey our wishes or our fears. Except on weekends, mathematicians must accept what Olympus reveals—or conceals.

The question that lurks behind the two questions I have posed always makes the hair on the back of my neck stand up: Could it be that the independent existence of mathematics has something to do with its power to describe the physical world with such accuracy? Think of it: At a purely mental level, all of mathematics lies before us—some known, a great deal yet to be discovered. Its truths are adamantine, incontrovertible. At another level, we live in a cosmos that apparently obeys mathematical laws. Why, in heaven's name, should that be?

I deeply hope that some of the scholars I have arranged to meet will provide answers to these questions, or at least clues. In my quest, I must leave no stone unturned, overlooking no possible explanation for what appears to be the independent existence of mathematics or its ubiquity in the cosmos. I must even be prepared for the possibility that mathematics is, after all, created; that I have merely been deceiving myself all this time. Whatever insight brings me to such a conclusion will surely arise from culture or history. For example, I may discover that Greek mathematics was influenced by early Greek culture to an extent that would make it impossible or highly unlikely that such ideas would be produced by any other culture. In this case, the independent existence of mathematics will be an illusion, fostered by a willing submission to culturally determined modes of thought from one age to the next. Its wide applicability will be a massive self-deception, hinging on a worldview that is shaped by the same modes of thought. This is a possibility that I must face, however reluctantly.

I am therefore determined to ask my questions of each of the scholars I will visit. My first meeting will be in two days, at an archaeological site on the coast of Asia Minor, now modern Turkey. Petros Pygonopolis is a historian of mathematics and science at

Athens University and a leading exponent of ancient Greek mathematics. He wants to meet me at Miletus, the ancient trading city where Pythagoras, the preeminent philosopher and mathematician of the early Greek period, spent much of his life. In this setting, I hope, Pygonopolis can throw some light on Greek mathematics and science as a whole. How better can we study the influence of culture on mathematics than in a world so vastly different from our own, so removed in space and time?

Next on my itinerary is Egypt—rather the southern desert of Jordan—where I will meet Jusuf al-Flayli, an astronomer at Cairo University. For years, he has pursued research into early Arab astronomy, more or less as a sideline, it must be admitted. Based on my e-mail and other correspondence with him, al-Flayli seems especially interested in the relationship between the mathematics and astronomy practiced not only by the Arab astronomers of the Islamic period but also by the preceding Babylonian, Indian, and Ptolemaic (late Greek) scientists. He has promised to take me on a miniexpedition to the desert where he intends to reconstruct what he calls the ancient perception of the universe. He has also offered to explain the paradigm shifts that have taken place since those early days, including the Copernican revolution. Again, I hope that al-Flayli will help me disentangle the threads of culture from whatever hard realities underlay early astronomy or the mathematics that made it possible.

Further travel will take me to Venice, where I have an appointment with Maria Canzoni, a former research physicist at the giant CERN Laboratory in Geneva. At present, she teaches the history of science at the Universita Ca Foscari di Venezia. Canzoni came to my attention on the Internet as the proponent of what she chooses to call modern Platonism. She claims that Platonism (in the restricted form I have already described) is not only possible in the modern world but also necessary to a full understanding, philosophically speaking, of the relationship between physics and the cosmos it claims to describe. Mathematics is the key to this relationship. She seems to have no doubt about its preexistent reality.

Finally, I will journey to England, where I have arranged to meet that icon of twentieth-century mathematics, Sir John Brainard, at Oxford University. I was extraordinarily lucky to get this meeting, not only because of Brainard's preeminence but also because of his advanced age. Of all my coming contacts, I know the least about this gentleman. Despite his expertise on the theory of computation (among other things), Brainard refuses to use e-mail. I have in hand just one rather cryptic letter from him, which promises to set me straight in the matter of mathematics and its relation to reality. He sounds rather crusty, and I look forward to our meeting with mixed feelings. Nevertheless, it is with Brainard that I have the highest hopes of settling my questions. He is said to be the last living mathematician who has a grasp of the whole of mathematics, if such a thing is possible.

MESDAMES ET MESSIEURS, NOUS ALLONS
DANS QUELQUES INSTANTS COMMENÇER
L'EMBARQUEMENT DU VOL AIR FRANCE 378
À DESTINATION D'ATHENE . . .

I will fly on the wings of mathematics first to the east, then back to the west—but how will I fly if I discover that there is no good reason for the plane to remain aloft?

PART I

THE HOLOS

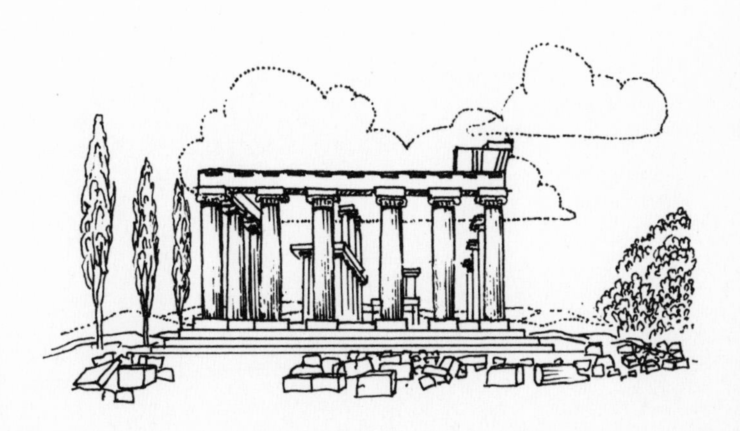

CHAPTER 1

Death of a Dream

Izmir, Turkey, June 22, 1995

A strange day! I spent it in Miletus, or what used to be Miletus in 500 B.C. It was there, if anywhere, that mathematics became a science. Miletus was the unrivaled center of commerce, of philosophy, and of the arts. Here lived Thales, the first scientist; Anaximander, the philosopher; and Timotheus, the poet. Here, too, the great Pythagoras visited from his native Samos, to learn and to teach.

First, let me backtrack, however. I arrived yesterday in Athens and changed planes for a one-hour flight across the Aegean, which took me to Izmir, Turkey, an hour's drive north of Miletus. This morning, in Izmir, I rented a Fiat Uno and drove south through a succession of beautiful valleys in supernaturally hot weather. I came at last to the Aegean coast, following it through relentless humidity until I arrived at Miletus, a collection of ruins with numerous tourist signs and a weed-filled parking lot. The site of the ancient city has been largely lost to silting and erosion by the Meander, the prototype of every river that twists and turns. I parked outside a fenced-off area and made my way through partial restorations of the ancient city to the temple of Apollo. There, some tourists were gathering their bags and cameras, departing singly and in groups for a waiting bus.

There are times, when visiting old places, that hints of an ancient presence overwhelm you, like ghosts, in broad daylight. You cannot be with a tour group to have this feeling, you must be alone. It came over me as I stood before the temple of Apollo Delphineus, its steps and pillars haunted by memories not my own. I looked around for Petros Pygonopolis, the man I was supposed to meet, but there was no one in sight.

As I ascended the steps to the temple floor, I saw a man kneeling as if in prayer, hunched over the perfectly fitted, square paving stones. As I quietly approached, I could see that he was measuring the stones with a bronze ruler. He was a large man, with a shock of black hair, gray at the temples, and an olive complexion. He looked strangely out of place, for he wore an elegant white suit. When I cleared my throat, he looked up, startled. His stared at me confusedly from beneath bushy brows, then retrieved a pair of glasses, which he put on. His face was suddenly wreathed in a generous smile. Who could it be but Pygonopolis? He rose to dust the knees of his trousers, then bowed.

"You must be Dewdney! Excuse me for not seeing you. Thank the gods those tourists have finally gone. This is not the thing to wear for fieldwork, no? Of course not!" he answered his own question. "I am Petros Pygonopolis, historian of science and specialist in Greek mathematics—that is, ancient Greek mathematics. "

We shook hands and stepped back to examine each another.

"Welcome to Miletus, and welcome to your intellectual roots," Pygonopolis continued. "The questions you have asked in your letter are the right ones, in my humble view. Is mathematics discovered or is it invented? Does it have an independent existence? How refreshing that people can still ask such questions! The answers, such as they are, begin with what I was just doing, measuring these stones with this ruler." He held up the bronze strip. "It is the *pechya,* or cubit, that we Greeks once used."

While most personable and charming, Pygonopolis had a

nervous and agitated air, as though something crucial hinged on our meeting. Was it because few people took much interest in his work?

"That's a strange ruler," I said. "It has no marks on it."

"It has no marks," Pygonopolis explained, "because I am not measuring these stones in the usual way. I am not interested in the dimensions of the stones or of the building itself. I am simply curious to know in what unit the builders worked. If measurements come out even with this pechya ruler, then the builders must have used the pechya. If not, I will try some other unit." Pygonopolis glanced at an assortment of bronze rulers leaning against a restored pillar. The ancient Greeks, he explained, had no less than 20 different units of measure.

"Yet even while I satisfy my curiosity, I am following in the footsteps of the great Pythagoras himself." Leaving this mysterious remark hanging in the air, he brandished the ruler he was holding. "Let us see if the temple was based on the pechya. If it doesn't work, I will try next the pygon. I will start all over, so you can see what I am up to."

He strode to the rear of the temple, then knelt once again on the floor, ruler and pencil in hand. He laid the ruler on one of the great, square paving stones, its end flush with one side. The ruler stretched a little more than halfway across the stone. Pygonopolis made a small pencil mark on the stone at the front end of the ruler, then slid it expertly along its own length until the back end met the pencil line exactly.

The front end of the ruler now fell well beyond the crack between stones. I ventured aloud that perhaps the pechya was not the proper measure for this temple. Pygonopolis only grunted something about waiting and seeing. He made a new mark, shifted the ruler again, and seemed unperturbed when its front end stopped nowhere near the next crack. He proceeded to march the ruler along the row of stones toward the front of the temple, talking as he went.

"Sooner or later, if this is the right unit, the front of the ruler will again come even with one of the cracks. Of course, the pechya may be incorrect. . . . Well, well. What do you know!"

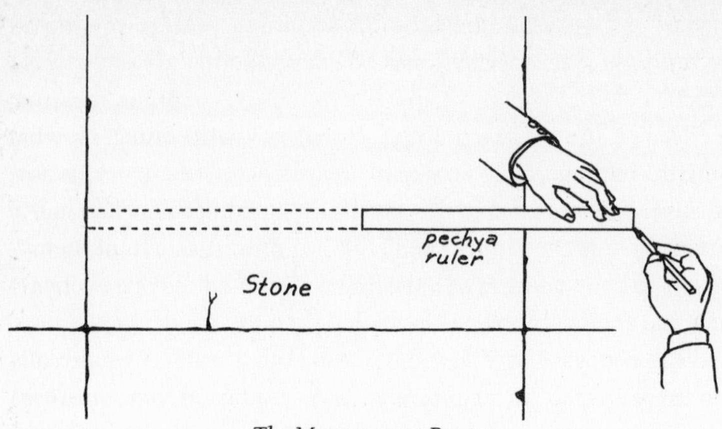

The Measurement Process

Near the portico, the front end of his ruler had met one of the cracks exactly. Excitedly, Pygonopolis took a small notebook from his breast pocket and made an entry.

"This was lucky," he commented. "There was no guarantee that the first ruler I should try would work. Let us see: In the process of measuring, of waiting until the ruler and the stones came even once again, I crossed 5 stones. At the same time, I measured 8 pechyas. From these facts we can deduce the exact size of the stones in pechyas, can we not?"

He stared at me expectantly. I would subsequently learn that whenever he ended a question this way, he expected me to answer it. Hastily, I set myself thinking. If 5 stones were, all together, 8 pechyas wide, then 1 stone must be $\frac{1}{5}$ of this distance, or $\frac{8}{5}$ of a pechya. I blurted the answer. "Eight fifths of a pechya or, if you prefer, $1\frac{3}{5}$ pechyas."

"Yes and no. I had better say something about ancient arithmetic. The classical Greeks had no sophisticated number system like ours. Their way of writing numbers symbolically was roughly equivalent to Roman numerals and not at all suited for calculation of any kind. Moreover, they had no way to express fractions such as $\frac{8}{5}$. Instead, they would speak of ratios of integers, such as 8-to-5.

"The important thing to notice here is that the measurement process came out even. With the ruler, I was measuring off an ever-increasing distance in pechyas. Meanwhile, I was traversing an ever-increasing distance in stones. Then, suddenly, the two distances coincided. Whenever this happens, you have a common measure, a certain length in which both measurements are integers, what some people call whole numbers. The whole numbers in this case are 8 and 5. The common measure is $\frac{1}{5}$ of a pechya. The pechya consists of 5 of these units, and each stone consists of 8 of them."

"So, did the builders of this temple work in fifths of a pechya?" I asked.

"It is entirely possible," stated Pygonopolis, "but what really interests me is not what unit the builders worked in, but something much deeper. In the end, it is not the pechya that counts but another measure, a special unit in which all measurements would come out as integers."

"I don't quite follow!" I interjected. I was becoming somewhat confused.

Pygonopolis suddenly leaned forward with a conspiratorial air. "It is entirely possible," he continued in a hushed voice, while he looked around him, "that the young Pythagoras himself stood in this very temple and measured these stones. He once did what I have just done. He was not determining the measure of the temple, either, but something far more profound." At this, the irrepressible Pygonopolis hurried me to the front of the temple where we could gaze at the Aegean Sea.

"Look over there!" He pointed to a long, mountainous island just across a strait to the west. "That is Samos, where Pythagoras was born, about 582 B.C." Pygonopolis swept his arms expansively along the strait. "At that time, this whole area, from north to south, was known as Ionia, a loose confederacy of Greek cities. Here, in the temple of Apollo Delphineus, we stand in the middle of Miletus, the most powerful city of Ionia, a center of trade and home to many philosophers in the real sense, men who interested themselves in everything. Here lived the great Thales, mathematician and teacher of the young Pythagoras. Thales was a merchant and a great traveler. From Egypt, from Arabia, and from the far Indus, he brought the mathematical riches that would become the foundations of Greek mathematics. And none was more influential in laying these foundations than Pythagoras himself. But there's much more to this story than mathematics, make no mistake!

"Somehow, perhaps through the influence of Thales, Pythagoras became convinced of an amazing doctrine, one that bears directly on your question concerning the independent existence of mathematics. Not only did mathematics have an independent existence, as far as Pythagoras was concerned, but it also had a powerful influence on existence itself, answering your second question. Pythagoras believed that what we call the real world was not merely measured by number, not merely described by number, but it was actually made of number—and, I might add, not just any numbers, but whole numbers, or integers. You could call it the integral universe. You could even call it a kind of digital universe.

"Can you imagine what this means? The whole idea is far more audacious than the timid doctrine of Democritus who, 100 years later, proposed a world made of atoms—hard, indivisible units. These were material units, after all, whereas the units Pythagoras proposed were immaterial, the integers. Can you imagine anything more immaterial than numbers? What a concept! Believe me, my friend, we are still catching up to Pythagoras."

These ideas swept around me, flooding me in a turbulent current. It was more than I had bargained for. There was also something of the impresario about Pygonopolis, something I could not wholly trust. We sat down on the temple steps, gazing out at Samos, while Pygonopolis caught his breath. Slowly, Miletus of old seemed to come alive around us, haunted by ideas that would never die.

"I have reason to believe that Pythagoras came here and to other places where he could experiment with, uh, commensurability. Ah, English! What an ugly word is *commensurability*. You know English, so you know what means this word, do you not?"

"Umm, let's see." I struggled to recall the definition. "Two lengths are commensurable if they have a common measure?"

"Just so. The pechya and one of those stones have commensurable lengths because they have a common measure, the ⅕ pechya."

I interrupted, "If you will permit a remark, most people see no need for such a difficult concept as commensurability because they think that any two lengths have a common measure, do they not?" (He had me doing it.)

"Just so. And for this they can surely be forgiven, for Pythagoras himself certainly thought so at one time. But I am getting ahead of myself.

"Commensurability is more easily grasped if you turn things around for a moment. Start with the unit. Suppose I have some unit, it doesn't matter which unit, perhaps it is very small. If I make two integer lengths out of this unit, any two integral lengths, the lengths will be commensurable. Suppose the lengths are 5 units and 8 units. If your ruler is 5 units long and the stones are 8 units wide, your measuring process is absolutely guaranteed to come out even, as it did when I measured the temple floor. As I moved the ruler into successive positions, I measured off an accumulating total length in fifths of a pechya:

5 10 15 20 25 30 35 40

"Now the widths of the stones were also adding up as I crossed them:

8 16 24 32 40

"You see, I arrived at a common number, 40. Sooner or later, the 5-unit ruler matched the 8-unit stones. The measuring process eventually came out even because the two lengths have a common unit. It had nothing to do with the integers themselves, as long as they *are* integers.

"This ruler is commensurable with those stones behind us because my measurements finally came out even. The connection is not obvious, of course. I will explain it later. The point is that the end of the ruler finally met a crack. Yet in a mathematical sense, there was no built-in guarantee that the ruler would ever do that, even if the temple floor extended infinitely! If the ruler ever meets a crack on an infinitely tiled floor, the two lengths, that of the ruler and that of the tile, are commensurable. They would turn out to have a common measure, like our ⅕ pechya.

"But we have been very sloppy. We must put some meat on the

bones of your definition by providing a precise test for the commensurability of two lengths. We will dispense with the stones altogether and speak instead of just two rulers. They are not real rulers, of course, just two strips of metal, each having a specific length. We will say that one of the rulers has length X and the other length Y. You may substitute any two specific measurements for X and Y that you like. What I am about to say will apply just as well to those two lengths.

X

Y

The Ruler Game

"This commensurability test is like a kind of game. We play the game with the two rulers. We begin by placing the back ends of the two rulers even with each other. We then slide the shorter of the two rulers ahead until its rear end comes exactly even to where its front end was. In fact, I have just given you the only rule of the game: Always take the ruler whose front end is behind and slide it forward by precisely its own length. That's it. The question is, will the two rulers ever come out even, their front ends matching? If they do, you win. The two lengths X and Y are, uh, commensurable. If the two rulers never come even, you lose. In such a case, the rulers are not commensurable."

"Hmmm," I grumbled. "You could end up playing forever, could you not?"

"Theoretically, of course, but we know this only as the beneficiaries of modern mathematics. We know that there are pairs of lengths for which the ruler game will never end, but Pythagoras did not know that. He knew, of course, that it was a theoretical possibility, but he believed that it would never happen. He believed that the world was structured in such a way that no matter what rulers you started with, you would always win the ruler game.

"As I mentioned earlier, the Pythagorean universe was based on integers. In practical terms, this meant that all lengths, whether of stones, rulers, or anything else, were ultimately integers. There was a fundamental unit in which all things would prove to have an integral measure. One possible test of this theory would be the ruler game. In such a world, it must always end in victory.

"This concept of a fundamental unit unified arithmetic and geometry in a particularly simple way. Arithmetic is about numbers, and geometry is about lengths. For every length, there was a privileged number, an integer, which expressed it. And every integer would, sooner or later, turn out to be the length of something or other.

"For Pythagoras, as well as for Thales and other early Greeks, arithmetic and geometry were already regarded as aspects of the same fundamental reality. A basket of figs always contained some definite number of figs, and a stone always had a definite size. Now the first kind of number was an integer. But what sort of number could one assign to the stone? Every ruler gave a different length, depending on the units it employed, and rarely did the dimension of a stone turn out to be an exact integer. It was far from obvious that there existed a privileged ruler, one marked off in these fundamental units I have been speaking of, by which the length of that stone, of all stones and everything else, would come out as integers."

Pygonopolis paused. "What I'm going to tell you now, you must listen carefully. Never mind the tape recorder. You will see all of Greek mathematics spin from this story, like the Golden Fleece of the sun.

"First I am going to show how Pythagoras would have proved the intimate connection between the ruler game and his integral universe. But that is hardly more than a sideshow compared to what follows. His integral universe collapsed when he discovered a pair of incommensurable lengths. For Pythagoras, it was a first-class crisis. A certain little diagram from Egypt had two lengths that, it could be proved, were not commensurable."

Clouds were gathering over Samos. Pygonopolis cast a worried glance at them.

"To begin with, what is the connection between the ruler game and the integer universe? Briefly, it is this. In the integral universe, you always won the ruler game. Conversely, if you always won the ruler game, you must be in an integral universe. It would not have taken Pythagoras long to prove that."

He had paused for breath again, so I interjected, "I'm curious to know how Pythagoras could have arrived at such a proof if the early Greeks had no algebra and could not even multiply or divide numbers, let alone symbols."

"We moderns might use X and Y to represent the unknown lengths, then apply algebra to prove the result. You are quite right to point out that the early Greeks did not have algebra, or an efficient number system. But they had something nearly as good when it came to proving results. As far as numbers are concerned, Pythagoras used a kind of symbolic geometry in which numbers were represented by configurations of dots. The configurations might be lines, triangles, or rectangles, all made of dots. For example, you could represent the number 10 by 10 dots in a row, by a rectangle 2 dots wide and 5 dots high, or even by the famed *tetractys* figure, a triangle with 4 dots on the base row, 3 dots in the next row, 2 in the next, and 1 on top, forming the apex of the triangle. Which representation you used for a number would depend on what you wanted to do with it.

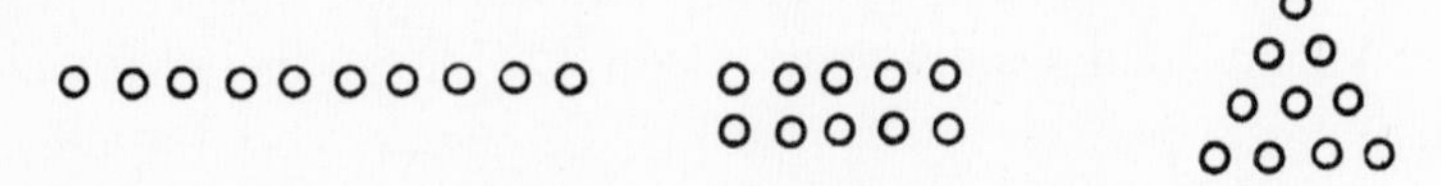

Dot Diagrams for the Number 10

"To represent algebraic ideas, such as ratios and products among unknown quantities, Pythagoras would use a geometrical figure, perhaps one that showed the successful outcome of the ruler game for two particular rulers. The diagram would show the positions taken up by the two rulers on their way to the final, successful outcome.

"By the way, I have no doubt that much of Greek mathematics, given its dependence on diagrams and geometry, was developed with Earth itself as the chalkboard. Archimedes is said to have been killed by a Roman soldier while pondering a problematic figure in the dirt. I sincerely hope there are no Roman soldiers about just now!" Pygonopolis sketched the following figure:

The Two Rulers Come Even

"To be sure, no one knows just how Pythagoras proved things. Only one thing is certain. The use of diagrams as part of formal proofs marks the singular success of Greek mathematics. It is a great strain to hold a detailed image of a problem in mind while pondering its components. To relieve the brain of this burden, the ancient Greeks learned to render the diagram with suitable precision in the dirt. This was a technological breakthrough of sorts. Their genius lay in applying geometrical thinking of one kind or another to these diagrams, replacing the algebra they did not possess by geometrical logic, which they did possess.

"Here is a case in point. How much easier it is to reason about the ruler game with such a diagram before one! Pythagoras would stare at it for at least a few minutes, mumbling to himself about the two lengths. Sooner or later, he would say Aha! He had found a proof that the ratio of the lengths of the long ruler to the short one was a ratio of two integers. From there it was but a short step to deducing the existence of a common measure, as we shall see."

"In his first and crucial step, Pythagoras would have matched each short ruler in the upper row with a corresponding long ruler in the lower one, noting that by the time he had counted his way to the end of the lower row, he would still be short of the end in the upper one—like so."

Pygonopolis marked off the corresponding rulers with *X*'s in the figure.

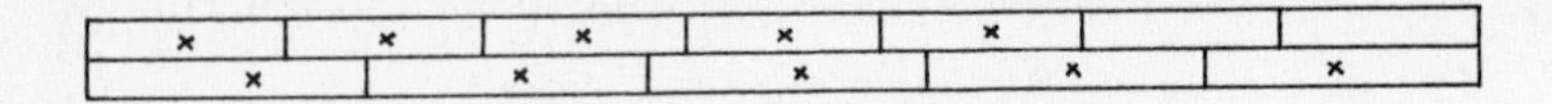

Matching Up Rulers in the Two Rows

"Now we have the lower row all crossed out and only a portion of the upper row so treated. However, because both rows contain the same number of crossed-out rulers, the ratio of the lengths of these rows must be the same as the ratio of the lengths of the rulers that compose them. Is that not so?"

It was, I said, perfectly clear. Dividing both integers of the ratio by this number would have no effect on the ratio. Although this was modern thinking, or seemed to be, I let the point pass. Presumably, the early Greeks had a geometric proof of this idea.

"Now, see how pretty this is! Pythagoras next imagined the long rulers of the upper row all shrinking until they had the length of the short rulers." Hastily, he drew a new figure in the dirt.

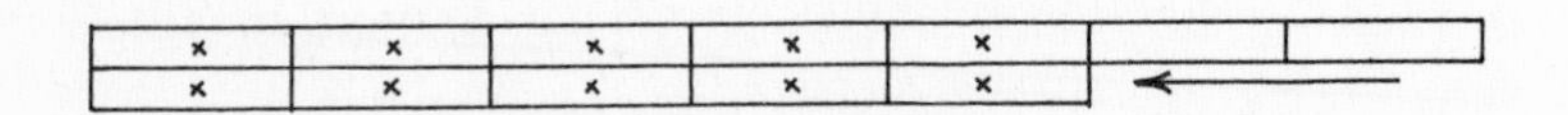

Shrinking the Longer Rulers

"Now you see what is going on, do you not?"

When I noticed that the shortened lower row had the same length as the crossed-out upper row in the previous diagram, I felt like saying "aha" myself. Mutely, I pointed to the two short rows, one in each diagram.

"Exactly. They are equal! In both cases, the ratio of the longer row to the shorter one is the same. In the previous figure we saw that this was simply the ratio of the long ruler's length to that of the short one. In the second figure, it is the ratio of the number of short rulers to the number of long ones. But these two numbers are integers. Therefore the ratio of the lengths of the two rulers is an integer ratio."

Pygonopolis had delivered the main proof, but there was something left. I pressed him to explain why the integer ratio meant the two lengths had a common measure.

"That part is the easiest. Simply divide the longer ruler into as many equal units as the larger of the two integers entering the ratio. Similarly, divide the shorter ruler into as many equal units as the smaller of the two integers. Because the ratio of the lengths equals the ratio of the number of units composing each ruler, the two kinds of units must be the same."

The proof had not been a difficult one, but my head spun a little, as if I had received a brain transplant. Early Greek mathematics had a very different feel from modern, algebraic reasoning. I ventured a question: "You have just shown me how Pythagoras might have proved the ruler game—that is, winning the ruler game was tantamount to the existence of a common measure. We in the modern age might proceed differently. We would work with the symbolic ratio X/Y and use algebra to prove the result. I feel a bit foolish asking this, but I must. Why should two entirely different systems of thought arrive at the same conclusion?"

Rather than becoming impatient with my question, as I feared he might, Pygonopolis looked pleased.

"This illustrates how two completely different trains of mathematical thought arrive at the same station, so to speak. It is a first-class phenomenon, when you think about it. Two completely different approaches to a problem, our modern algebraic approach and the old geometric one, lead to precisely the same result. Is it a coincidence? If you view mathematics as a purely cultural activity, you will miss a crucial point: It is not, in my view, a coincidence." Then he laughed.

"When some people talk about the cultural element in Greek mathematics, I fear they imagine Pythagoras dancing on the beach like Zorba, with a bouzouki playing in the background."

A brief rumble of distant thunder rolled across the strait from Samos, where clouds were gathering. Pygonopolis shivered slightly, staring down in silence at the diagrams. This was my opening.

"If it is not a coincidence, what is it?" I asked.

"It is essentially the phenomenon of independent discovery, the same idea finding a completely different expression by two people or groups of people separated by space, by time, or by culture. The phenomenon has been repeated thousands of times throughout the history of mathematics, and it points to something very special going on in mathematics. I suppose my own beliefs on this point are not very different from those of Pythagoras. For even after his integral universe was shattered, Pythagoras continued to believe that mathematics had an independent existence, although not in a material sense. But what, I ask, did he call it?

"Pythagoras was a *mystic* in the traditional sense—someone who practiced inner discipline to arrive at new levels of understanding. Perhaps I will say more about that tomorrow. In the meantime, I can tell you only my opinion: He surely had a name for the place where mathematics exists. I have tried to imagine what this name might be. My best guess is the Holos."

"The Holos?" I repeated, as this was an unfamiliar word.

"The Holos is the place of mathematics. It stands in a special relation to the cosmos. Holos the source, cosmos the manifestation."

Pygonopolis paused, breathless again. The new word echoed in my mind. The *holos*, the holos, a beautiful word, pronounced with the Greek letter chi, a rasping *H*, followed by an ululation.

"Earlier you described the Pythagorean universe of integers," I remarked, "but all along you've been hinting at a tragedy. What happened?"

"As I said before," he responded in a patient tone, "the major underpinning of the integer universe, as Pythagoras imagined it, was what we could call the hypothesis of cosmic commensurability: Every two lengths were commensurable, not just in practice, but also in principle. There can be little doubt that during the time that Pythagoras believed the hypothesis, he bent every effort to proving it. He worked geometrically, trying one approach after another, but all his efforts came to nothing. No matter how much he wanted the hypothesis to be true, he could not prove it. Nevertheless he continued to imagine that the integers, specifically the number one, was the *atomos* from which the gods made everything. Ah, what a blow it was!"

"What happened?" I asked.

"His supreme vision was shattered when Pythagoras found the first pair of incommensurable magnitudes. Perhaps it was his old teacher, Thales, who suggested that Pythagoras check the commensurability of the side of a square with its diagonal. See, here it is.

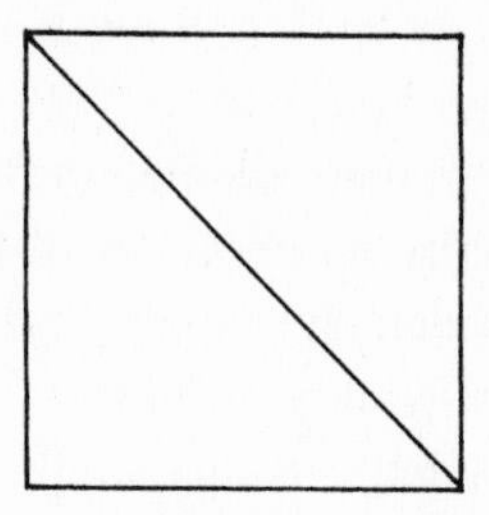

Square with Diagonal

"If the universe was based on integers, all pairs of lengths would be commensurable, including the two lengths in this innocent-looking little diagram. One of the lengths is the side of the square, all four sides having the same length. The only other length in the figure is that of the diagonal. It doesn't matter what size you draw the figure, as we are concerned only with the ratio of the two lengths. Was it an integer ratio or not?

"Pythagoras may have puzzled over this longer than he should have. Sometimes mathematicians are slow to discover the truth about a pet idea because they fondly imagine it to be true and are continually trying to find a proof of it. They never seriously seek to disprove it. But now Pythagoras had a test case to consider. How long did it take him to realize that it was what we call a counter-example?

"One day it came. The discovery staggered him, for it brought the integer-based cosmos to a huge nothing. Once he had gotten over his shock, he felt immense gratitude that at last the question of commensurability had been settled—in the negative, as it turned out. Up to this time, Greek mathematics recognized only two kinds of numbers, the integers and their ratios. Now there appeared to be a mysterious third kind of number, one that called for a revision in thinking. A new world had opened.

"Here's where the cultural element comes into play: His gratitude was so great, he went to a temple—perhaps this very one—and sacrificed an ox. We moderns do not understand sacrifice, by the way. Imagine feeling so grateful for some wonderful event that, to relieve your heart of its burden of joy, you buy a Mercedes and set it on fire!

"The argument that Pythagoras used to show the incommensurability of the side of a square with its diagonal is quite simple when you write it in modern symbolism, but we will prove it more or less in the way that Pythagoras did. We will not use algebra, then, but we will allow letter names for parts of the diagram. In particular, call the short side X and the long side Y. These, you will agree, are not algebraic variables. We begin with the very figure that Thales showed Pythagoras."

Pygonopolis stabbed Thales's figure with his stick.

"Thales had been to Egypt and had learned many wonderful things from the Egyptian priests, including this interesting little fact about the side of a square and its diagonal." Pygonopolis drew a second square tilted in relation to the first. One of the sides of the new square was the diagonal of the first one.

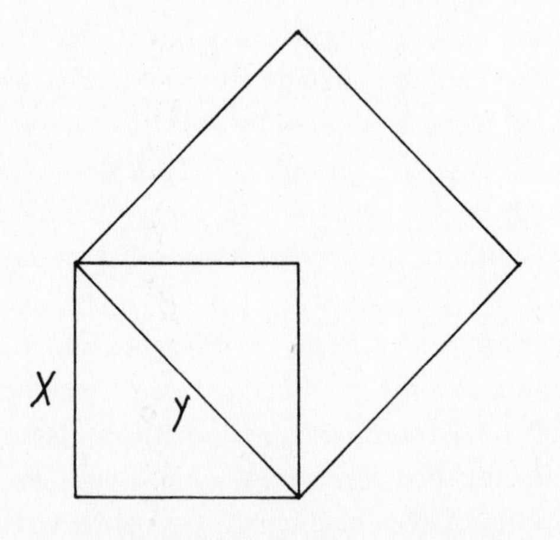

Square with Square on Its Diagonal

"The Egyptians, who labored under the same restrictions as the Greeks, had been clever enough to discover a curious relationship between the two squares. The larger one has twice the area of the smaller one. "The Egyptian proof was simple. You simply add 2 new lines, like so, and realize that the large square has been divided into 4 small triangles, while the small square is already divided into 2 of the same triangles: 4 is twice 2. *Quod erat demonstrandum,* as it says in the old texts."

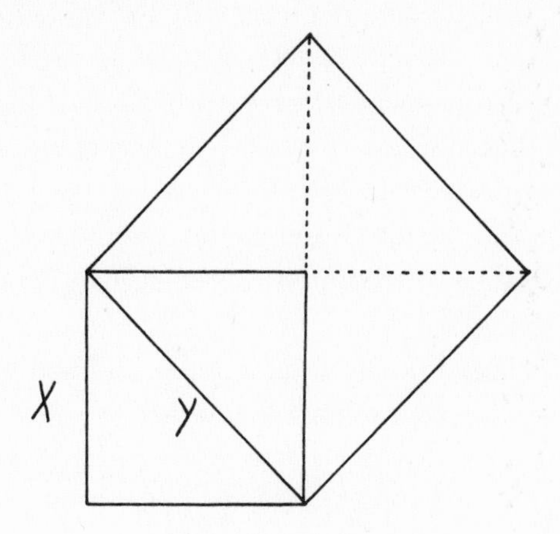

Egyptian Proof of Relation

Pygonopolis carefully smoothed away the two construction lines, restoring the earlier figure. Absentmindedly, he brushed his hand on his suit, then swore. "Agh," he said, "what a stupid thing to do. See what I've done!" He spent a few moments brushing the smudge on his suit, frowning.

"After a certain amount of the usual fumbling around that mathematicians go through, Pythagoras found the first step of his proof. If he assumed that X and Y were commensurable, then both X and Y had integer lengths in the unit of their commensurability. He also insisted that these numbers be the smallest ones having this property; this meant that the integers could not have a common factor.

"He could visualize not only the lines X and Y as rows of dots (the units) but also both squares made up of those dots. In particular, the number of dots in the large square was an even number, being twice the number of dots in the small square. Pythagoras then asked himself, 'Can an odd number be squared to produce an even one?'"

"My dear Professor Pygonopolis," I interjected. "I thought you said that Pythagoras had no algebra, and I assumed that meant no squaring."

"No, no, no, Dewdney. As I explained earlier, the ancient Greeks could multiply by geometry, and this meant the operation of squaring, as well. In this case, he drew a number as a row of dots. To square the number geometrically, he made a square of it, literally. He added more rows of dots above the first, as many as it took to produce a square shape. In fact, this is where the English word *squaring* comes from. In any case, the total number of dots in the square is the product of the number of dots along the bottom and along one vertical side.

"Pythagoras undoubtedly already knew, and had proved, that when you squared an odd number in this manner, the total number

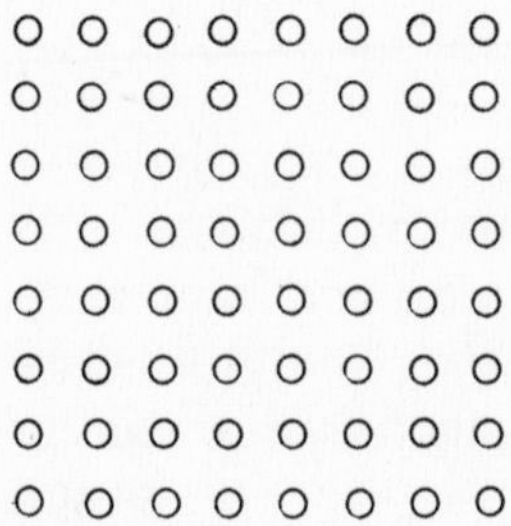

Squaring a Number Geometrically

of dots in the square was always odd. And, when you squared an even number, the result was always even. Now, according to the Egyptians, the larger square had twice the area of the smaller one. This meant that its area, the number of dots in it, was even. But, as we have already seen, this could only have been the case if the length of the side being squared—namely *Y*—was also an even number.

"Now the pace picks up. If *Y* had an even number of dots, then its square would have not merely a twofold (even) number of dots but a fourfold one. In modern language, this means that the square of *Y* was a multiple of 4.

"Now recall the Egyptian theorem: The square of *Y* was twice the square of *X*. Yet the square of *Y* is also a fourfold number, a multiple of 4. This meant that the square of *X* must have been a twofold number, or multiple of 2. You see where this is going, do you not?"

"Are you going to apply the same reasoning all over again to *X*?" I guessed.

"Exactly. Pythagoras could now apply the same reasoning to *X* as he had to *Y*, concluding finally that both lengths consisted of an even number of the fundamental units that made them up. This meant that if you cut each of the two integers in half, you would get new, smaller integers with the same property: Their ratio would be *X* to *Y* again. Because the integers in question were already the smallest possible, however, this was a contradiction. The logic refused to cooperate further. The machine had ground to a halt. In such cases, the Greek mathematicians, no less than we moderns, knew that one of the assumptions going into the proof must be wrong. There was only one assumption made—that the lengths of *X* and of *Y* were commensurable. The contradiction meant that they couldn't be."

Pygonopolis sighed, seemingly to catch his breath. "Can you imagine? Can you just imagine this moment for Pythagoras? There was no doubting the new result. Instead of proving the long-sought theorem, 'Every pair of lengths is commensurable,' he had proved exactly the reverse: 'There exists a pair of incommensurable lengths.' Although it doomed his doctrine, at least in its existing form, I dare say that Pythagoras was secretly delighted. He sensed higher ground ahead, as though scaling Olympus itself. The numerical *atomos* was deeper and more complicated than he had thought. There was another kind of number lurking in the holos and, therefore, in the cosmos. It was not an integer, nor was it a ratio of integers. We moderns call such numbers irrationals, meaning only that they are not rational numbers."

Storm clouds were gathering across the strait with Samos. Thunder rolled more and more frequently across the sea.

It was now late afternoon. Pygonopolis picked up his hat and strolled into the temple. I stood in a trance, watching him collect his briefcase, then stride out to the steps again, smiling broadly.

"I feel as though I have overwhelmed you in some way. It is exciting, this story of Pythagoras and the incommensurables. It shows so many things about early mathematics, yet we only have to stretch our imagination a little to understand how the early mathematicians must have worked."

I nodded as we parted for our separate cars. The rain fell heavily as we each drove north to Izmir. I had been hoping to ponder the things I had learned today as I drove but had no calm moments for reflection. Instead, I was hard put to keep up with Pygonopolis's rented Mercedes. He drove like a maniac, in spite of the bad road conditions. I was vastly relieved when we reached the outskirts of Izmir.

We had supper this night, Pygonopolis and I. We dined on Mediterranean seafood in the hotel. By asking Pygonopolis about the holos, eating as he talked, I succeeded in finishing long before he did.

"But just what is the holos, do you think?" I asked.

"The holos is the place where all of mathematics, both known and as yet unknown, exists," he responded cheerfully. "Here are the definitions, the axioms, the rules of deduction, the theorems, and the proofs. Here, too, are all the numbers, number systems, sets, families of sets, and on and on and on."

"But what I really want to know," I pushed my point, "is how such things can exist. Do they exist independently, like this chair?"

"The answer is subtle," he said, "for the existence of these things seems to depend on a human mind, yet it doesn't. Think of the number 3, for example. Wherever there are 3 of anything, the number 3 is also present, and not just as a concept.

"Here, for example, are 3 prawns left on my plate. Just 3. The threeness of these prawns will control how many more prawns I can eat without ordering more. In short, I cannot eat more prawns than are on my plate. Not only is the threeness of the prawns manifest to our senses and minds, but it also has an operational significance that goes beyond my conception of threeness. My inability to eat more than 3 prawns from my plate has nothing to do with my conception of threeness, or even the fact that it is me sitting here. Anyone else would face the same options. Likewise, if I order 10 more prawns, I may calculate that the number of prawns on my plate will then be 13. In this and in many other ways both simple and complex does mathematics control the world. In this way do holos and cosmos interlock."

"If I understand you correctly," I followed up, "the holos is a real place, although not in our ordinary sense of being locatable in our universe, or cosmos. Yet it also controls, to some degree at least, what goes on in our world, in the cosmos. What I'm not clear about is how much of this theory is due to Petros Pygonopolis and how much to Pythagoras."

"The theory of the holos, as I describe it, is entirely my own. An extended fantasy, if you like. Yet I cannot help but believe that Pythagoras thought of mathematics in pretty much those terms. He had seen how numbers evanesce when you consider what it is that unites all collections of 3 things. He had also seen how lines vanish as they are drawn ever more accurately, finely incised on smooth, flat stones. He had witnessed then, as anyone can today, how these concepts retreat when they are pursued, as though fleeing back to the holos. Yet they advance to take control at other times."

"This afternoon you made a joke about Pythagoras dancing to bouzouki music on the beach." I had not understood this earlier reference. "I assume he didn't actually do this, so what role did early Greek culture actually play in Greek mathematics?"

"Let me make an analogy. Mathematics is like the wheel. Almost every culture has its wheel, and the wheels made by different cultures all look different. An Egyptian chariot wheel is quite different from that on a medieval European oxcart, and both wheels are again different from those on a modern automobile. Yet all wheels operate on exactly the same principles.

"Nevertheless, cultural clues are crucial to understanding Greek mathematics—not so much its validity or universality as its direction. On one hand, you will find nothing discovered by my forebears that could not have been discovered by a Solomon Islander. But you might also find that the Solomon Islander would not be very interested in Pythagoras's problems, so he would be very unlikely to pursue them. It is early Greek culture that molded Pythagoras's mind. At the center of this culture resided the gods. He accepted them as perfectly real, and his deepest ponderings on the ultimate nature of the cosmos necessarily included the gods. He believed in the idea of a controlling presence such as Zeus, assisted by other presences in the pantheon. But this was the horse that pulled his cart and not the cart itself, so to speak. It motivated his search, even inspired it, but it did not play any role in what Pythagoras actually discovered—unless of course . . . "

Pygonopolis had begun to perspire heavily after finishing his meal. He paused to pass a handkerchief over his face.

"Let us just say that as far as Pythagoras was concerned, the controlling presence had left clues about itself, and Pythagoras, eager to scale Olympus, cast himself in the role of cultural hero. He saw in mathematics the path to knowledge such as only gods enjoy. Logical consequences resided in the very fabric of existence. It was surely how the gods *worked*."

Pygonopolis stared intently at me with his dark brown eyes. The hair on my neck stood up, and I felt, just for a moment, that he knew far more than he was telling. It occurred to me that he, Pygonopolis, actually believed in the gods of old—but then, abruptly, the mood passed.

"Tomorrow morning, we will take the plane for Athens. It is late."

Pygonopolis looked at his watch and, in so doing, revealed the palm of his left hand. There was a tattoo there—a small blue star. I glanced away quickly as Pygonopolis looked back at me from the watch.

As we parted, be observed, "You cannot know what it means to me to have a real listener."

CHAPTER 2

The Birth of a Theorem

Early in the morning, Pygonopolis and I took a taxi-bus to the airport and boarded the plane for Athens. The aircraft climbed out over the Aegean, the island of Samos slipping under our wing to the south. The sun was brilliant but only deepened the blue of the sea below. Pygonopolis nudged me at the sight of an oil tanker, seemingly immobile, trailing a pale wake.

"You must imagine the Mediterranean 2,500 years ago," he said. "No boats that size, but small sailing vessels you would hardly notice from this height. Ah, what days those were! Greeks, Egyptians, Phoenicians, Berbers, and the rest. The Mediterranean was the center of a heroic world, the Homeric age, when the gods ruled. From this height, we would hardly notice the small ship taking Pythagoras from Miletus to Crotona, a Greek colony in Italy, where he would found a school devoted to reason and mystery."

"Mystery?" I asked.

"Yes, something like the Eleusinian mysteries, an occult school of carefully selected students who could carry on the work of

Pythagoras when he died. Later today, if the gods are kind [and here he winked at me to let me know he wasn't entirely serious], we will be in my office at Athens University and I will tell you the story of Pythagoras's greatest achievement. In the meantime, I left some loose ends from last night."

"What loose ends?" I asked.

"About being wrong and about numbers. I was reflecting, just before I fell asleep, how this business of being wrong, of the possibility of being wrong stalking our every step, is what gives point to mathematical research. Not everyone understands what Francis Bacon meant when he said, 'Truth proceedeth more readily from error than from confusion.' In other words, it is better to be working on a hypothesis that turns out to be wrong than to have no hypothesis at all."

"Do you think that Pythagoras understood that his idea of the cosmos being founded on the integers might be wrong?" I asked.

Pygonopolis went rigid in his seat, a kind of mock astonishment.

"Of course. We Greeks invented the word *hypothesis*. *Hypo* means 'lying under,' and *thesis* means 'idea' or 'theory,' at least in this context. It is a foundation that one must test before building anything upon it. The little diagram from Egypt blew away this foundation. It forced Pythagoras to abandon his hypothesis. On the other hand, he was free to modify it and may well have done so. We are not privy to this particular development. However, since he and the brotherhood he subsequently founded continued to take number as the basis of reality, he may well have found room for the new numbers. After all, they promised to complete the link between geometry and arithmetic, which brings me to the second loose end.

"It is ironic that those straight lines the early mathematicians drew to compose their figures incorporated the new numbers. They knew very well that if they fixed a point on a straight line, then measured in one direction, every point on that line would lie at a certain distance from the fixed point. Until the disaster of the incommensurables, Pythagoras would have said that all the distances were rational, therefore all the points on the line corresponded to rational numbers.

"Now if one takes the side of the Egyptian square as a unit distance in some scale, then the area of the square will also be a unit, or 1. But, as you may recall from yesterday's discussion, the square on the diagonal was twice this area, or 2. This meant that the length of the diagonal, when squared, was 2. Pythagoras therefore knew that this new incommensurable magnitude, the one that could not be reconciled with any integer or rational number, was the square root of 2. I find it interesting that Pythagoras called the new kind of number αλογοσ [a-logos] or 'illogical,' whereas we call it *irrational*, meaning 'not a ratio.'

"The root of 2 was not a rational number, and this, in turn, could only mean that their understanding of such a simple thing as a straight line was woefully incomplete. Any straight line made entirely out of rational numbers would have an infinitesimal gap where the square root of 2 was supposed to be. What if there were many other illogical numbers, as well? Not to mention completely new kinds of illogical numbers. It was not until the nineteenth century that we had a complete picture of the so-called continuum, or straight line. As it turned out, the irrationals were a whole order of infinity more numerous than the rationals. Yet we also discovered that there were no more kinds of numbers to discover in the continuum—nothing beside integers, rationals, and irrationals."

"Did Pythagoras really regard irrational numbers as illogical?" I asked.

"That is a poor translation. He arrived at them by a process of logic, of course, so they were not, in fact, illogical. *Alogos* means something more like outside of the word or beyond the pale, as you English say."

We flew to Athens together, Pygonopolis and I, then drove to Athens University, passing the high hill dominated by the Parthenon. Lining the street below the ancient temple and center of old Athens were prostitutes, male and female. Pygonopolis clucked his tongue, then suddenly burst into laughter. "That reminds me of a joke. What is the world's second oldest profession?"

I confessed to not knowing. "Mathematics, of course!" Pygonopolis slapped his knee. When he had recovered, he asked,

"In what other field of human inquiry do we find results 2,500 years old that are still in active use today?" I nodded as he parked the car.

His office overlooked a delightful courtyard of olive and fig trees. "I would go down and draw figures in the dirt," said Pygonopolis, "but the chalkboard is more convenient.

"I am in a position now to complete the story of Pythagoras. Yesterday we witnessed the death of theory. Today I will tell you about the birth of a theorem, his greatest. You know the one I mean, I think."

Pygonopolis drew a right-angle triangle on his chalkboard, labeling the sides of the triangle *A*, *B*, and *C*.

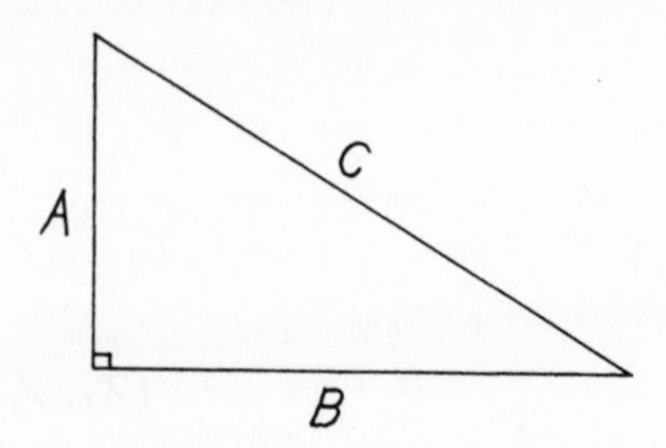

The Pythagorean Theorem Begins

Next, he drew a square on each of the sides, then he marked the side labeled *C* the "hypotenuse."

"The Pythagorean theorem, as you know, tells us that when you square the lengths of the three sides of a right-angle triangle, a certain relationship holds among the squares. The square of the hypotenuse equals the sum of the squares on the other two sides. Algebraically, we write,

$$C^2 = A^2 + B^2$$

"It is a strange theorem, one that creeps up on the newly introduced. If I tell such a person that the area of the large square on the long side of the figure equals the sum of the areas of the other two squares, they will certainly be tempted to say, 'So what?' And to this I reply that the statement is true for every possible right-angle triangle without exception. And to this I add that the truth of the statement is not obvious. Great Zeus! Why on earth should the squares on the three sides have this particular relationship? Why not some other? Why any?

"And if this does not impress them, I say not only that the theorem is true, but also that it lies at the foundation of every geometrical subject and is essential to a myriad of calculations that we make every day. For example, we locate points in physical space by their coordinates, say *X* and *Y*, for the sake of argument. Here." Pygonopolis added coordinates to the figure on the board.

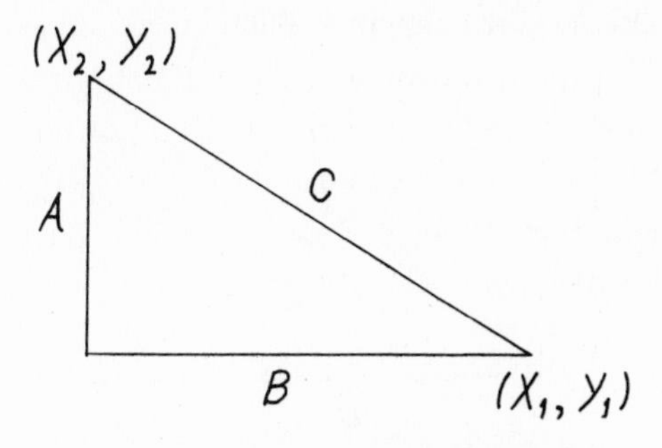

The Theorem Has Many Applications

"Such coordinates, like those on a flat map, locate every point by means of two numbers, as measured from a fixed point, either on or off the map. There is a horizontal measurement, the *X* coordinate, and a vertical one, the *Y* coordinate. What are the lengths of the sides *A* and *B*? Why, they are merely the differences $(X_1 - X_2)$ and $(Y_1 - Y_2)$. Now square these lengths and add them. You get the square of the distance between the two points. Now take the square root of that, and you have the actual distance. This calculation is repeated millions of times every day in guidance computers on aircraft, ships, and spacecraft.

"Come to think of it, the theorem illustrates what you have called the 'unreasonable power' of mathematics. If the Pythagorean theorem were not true in a most essential sense, aircraft would crash, ships would run aground, and spacecraft would be lost forever. Why should that be? Because space itself has the properties assumed by the theorem. I could say much more about this subject, but I am afraid of getting sidetracked."

"By all means, let us hear about the birth of a theorem," I interjected. "Tell me about the applications later, if you would be so kind."

"How did Pythagoras find this amazing theorem? You will remember that his digital world collapsed with the discovery of incommensurable numbers. As I hinted yesterday, Pythagoras was secretly delighted. He sensed higher ground ahead of him, as though he were scaling Olympus itself. There was another kind of number lurking in geometry and, therefore, in the world.

"Where else to probe this mysterious new universe but in the figure that had proved so troublesome in the first place? I refer to that tricky little square from Egypt." Pygonopolis drew the original figure of the square with its diagonal, brushed away half of it to reveal a right-angle triangle, then labeled the diagonal as C and the other two sides A and B.

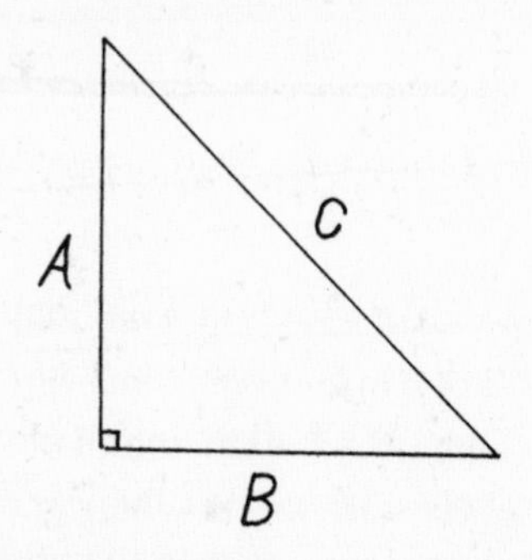

The First Example of the Theorem

"In the proof of yesterday, we saw that $C^2 = 2A^2$. This was the heart of the argument showing that A and C could not be commensurable. How just, how fitting, that the source of the problem, this accursed Egyptian disease, should be made itself to cough up an entirely new result by way of compensation. For the theorem that Pythagoras suspected to be true already lurked in this diagram. He could see that C^2 equals A^2 plus B^2. In other words, the square on the hypotenuse equals the sum of the squares on the other two sides. Here, the two sides *A* and *B* are equal. This was essential to the theorem, was it not?"

He didn't wait for me to answer, plunging recklessly ahead.

"Of course not! The same thing was true of other right-angle triangles. For example, the Egyptians had known for centuries of a special figure called the 3–4–5 triangle. Egyptian builders evidently used this triangle as a ready source of right angles. They used a long rope, its ends knotted neatly together. Two other knots in the rope

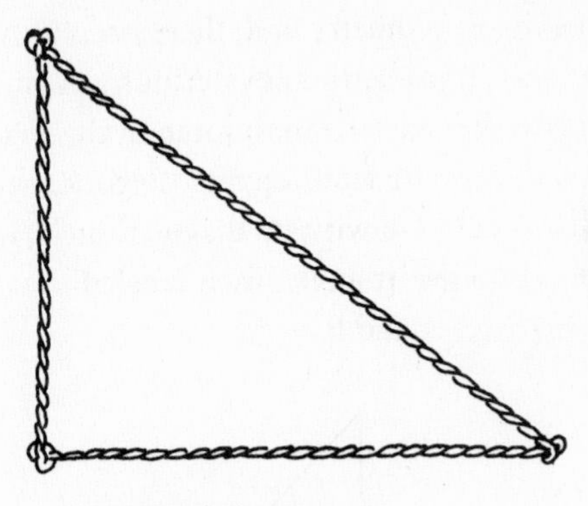

The Egyptian Rope Trick

divided it into three sections altogether, and the lengths of these sections were 3, 4, and 5 cubits, say. Now when they stretched this rope out tight, placing a peg at each knot, Lo! A right-angle triangle would appear. With such an instrument, right angles good enough for buildings and monuments could be laid.

"It was undoubtedly Thales who told Pythagoras of this trick. How could either of them have failed to notice that the three sides had a special relationship when you squared them?

$$3^2 + 4^2 = 5^2$$

"And how could Pythagoras fail to suspect that the same thing held for other right-angle triangles? He expanded his study at first to the right-angle triangles in which all the sides have integer lengths, whole-number lengths, like the 3–4–5 triangle. Today we call these Pythagorean triangles. He probably gave them a different name. I will call them *atomagonos* or, for you, *atomagons*. The sides of these triangles, unlike the sides of the troublesome triangle, all had integer lengths and were therefore all commensurable—no square roots of 2 lurking anywhere. Pythagoras knew that if the theorem were true of all right-angle triangles, with commensurable sides or not, it would be true of all atomagons. He hoped that the path to a proof might lead through the atomagons. He was not wrong, as it turned out.

"There can be little doubt that Pythagoras looked at quite a few atomagons as he struggled toward the great theorem that today bears his name. With computers, it is not difficult to write a simple program that will generate as many atomagons as you wish. Some of these were undoubtedly found by Pythagoras, as well. How could he not? They were already in the holos, waiting to be found. Here are all the atomagons with sides of length 25 or less." He handed the paper to me.

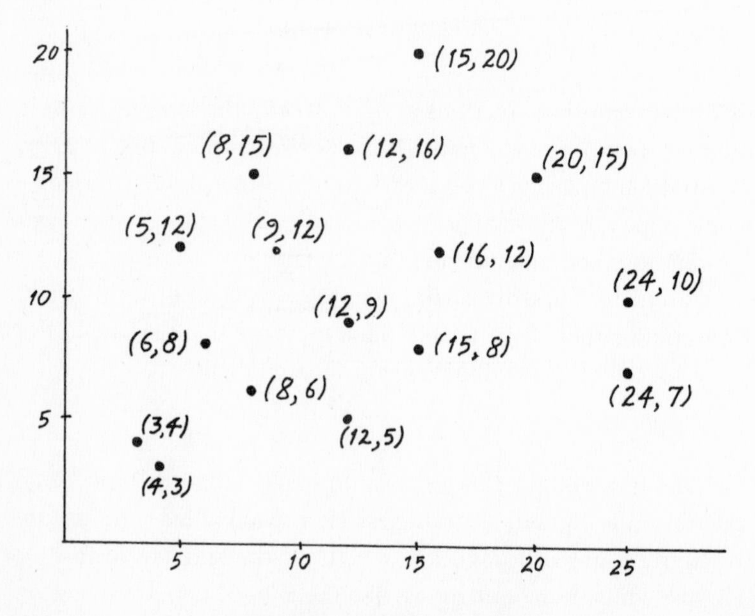

A Graphical Table of Atomagons

I looked at the table carefully. There were no triangles, as one might have expected, but pairs of numbers instead. It took me a few moments to realize that the pair (3, 4) actually meant the 3–4–5 triangle, and each of the other pairs of numbers represented the lengths of two sides of a right-angle triangle. To discover the hypotenuse in each case, you had only to square the numbers in the pair, add them, then take the square root. The two numbers in the automagon (65, 72), for example, produced the sum

$$65^2 + 72^2 = 9,409$$

and 9,409 turned out to be the square of 97. This was the length of the hypotenuse.

"But this is nothing, marvelous as computers are. Pythagoras found a way to generate atomagons of any size without using a computer. The method depended on a certain highly significant shape called a *gnomon*, the old Greek name for a carpenter's square, a flat instrument consisting of two rectangular strips joined at right angles, like a corner bracket.

"To construct one of his atomagons, Pythagoras would start with a square of side A, an integer. He would then fit a gnomon that was 1 unit wide onto the corner of the square, thereby producing a slightly larger square of side $A + 1$. For example, if side A was 4, he would take a gnomon of width 1 and fit it to the corner of the square. See, here it is on the board.

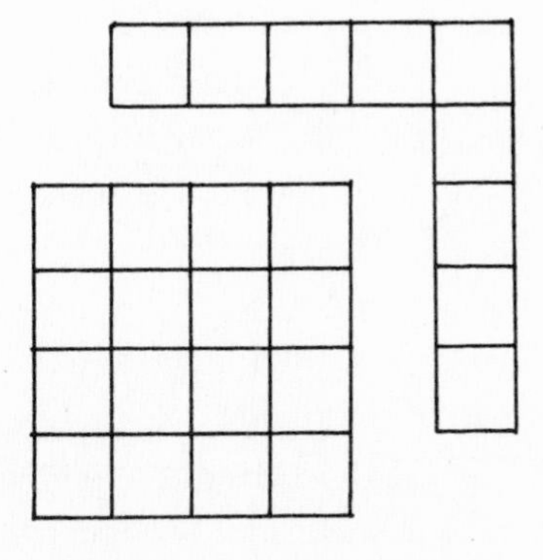

Adding a Gnomon to a Square

"The area of the gnomon, the number of little square units in it, was odd, was it not? It would consist of two equal arms with a combined length that was even, with the little unit in the corner added. If it so happened that this odd number was also a square, such as the 9 in my figure, he would have an odd number of the form B^2, and then he was done. For then he would have three square numbers, two from geometrical squares, the third arising from the gnomon: The first square number came from the geometrical square he started with, the one of side A, and the second square number came from the gnomon, not itself a square, but having a square number of units. The third square number came from the geometrical square obtained by adding the gnomon to the first square. It would certainly be true, would it not, that the sum of the first two squares would equal the third?"

"Yes it would," I answered, "but how on earth did he find the values B for which this scheme worked?"

"Pythagoras merely had to count his way through the odd numbers: 1, 3, 5, and so on. Every time he came to a square number, he would have a new atomagon. Here, try it. Is 7 a square? No. Is 9? Yes, indeed. The atomagon here would have sides 3, 4, and 5, our old Egyptian rope trick. Of course, this particular method suffered from a limitation. The gnomon was always a strip of width 1, so that the atomagons produced always had two sides that differed by only 1 unit. I must add, without going into it here, that Pythagoras soon found a way to extend his method so that he could finally generate all the atomagons.

"His exploration of the atomagons complete, Pythagoras had established the theorem for all integer-sided right-angle triangles. Perhaps he sacrificed another ox. He realized, thanks to the Egyptian figure Thales had shown him, that there were right-angle triangles that were not atomagons. The question was, could his unveiling of the atomagons provide the key insight he needed to complete the theorem of the squares? If so, where was it? Here is his favorite diagram of the time, the one to which his intuition had guided him. He may have stared at it for hours and hours, haunted by the feeling that the solution was staring him in the face.

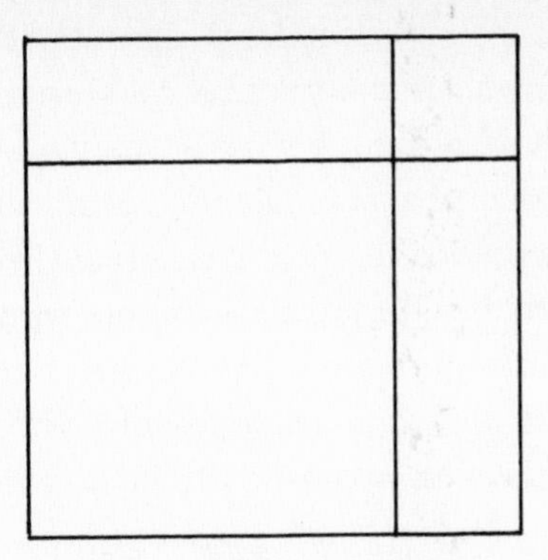

Square with Gnomon Added

"As you can see, he had changed the gnomon slightly, cutting off its corner into a separate square. The figure therefore consisted of a large square, a small square, and the rest of the gnomon. He sought a construction line or two, something that would bring the key insight out of hiding. One day, he drew this line through one of the arms of the gnomon.

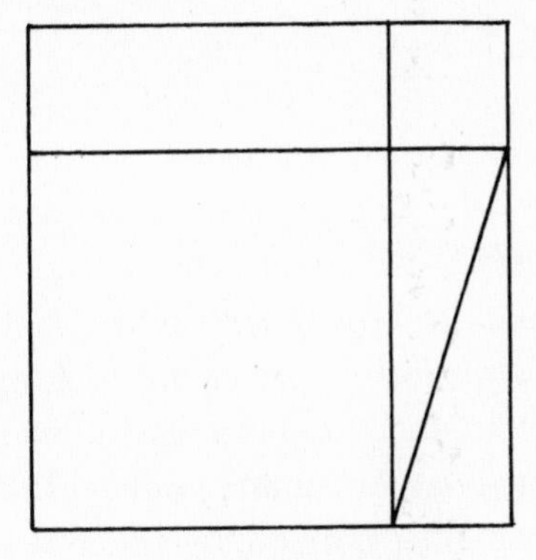

Gnomon with Diagonal

"This was the Aha! moment. It was the summit of Greek mathematics and a divine moment for Pythagoras. He saw that the arm of the newly bisected gnomon contained a right-angle triangle. Could this be the right-angle triangle from which his theorem might spring? Its smallest side belonged to the square in the corner of the gnomon. The larger side, at right angles to the smallest, belonged to the square embraced by the original gnomon. The hypotenuse—to what did it belong? Pythagoras had the boldness and insight to discover a new diagram, one in which the arm of the gnomon was distributed around the big square like so." Pygonopolis drew a new square like the previous one, but with four copies of the gnomon's arm distributed around the square. In the process, a new, tilted square had appeared inside the old one. "That!" he cried, stabbing the figure with his chalk, "That was IT!"

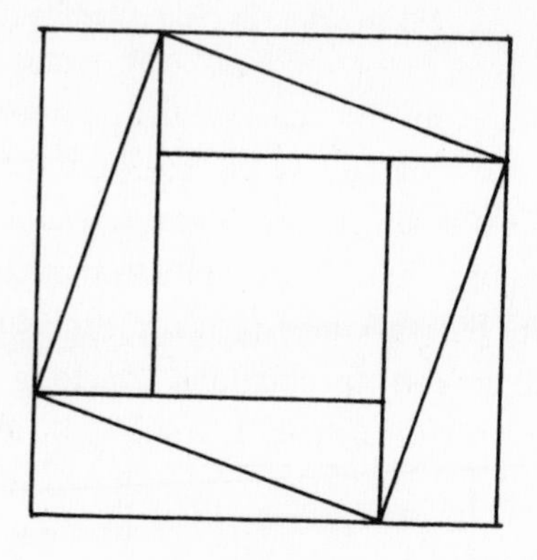

Three Squares

I could recall reading about this construction, but only now did its marvelous qualities become apparent. The tilted square produced by the construction was simply the square of the hypotenuse of the right-angle triangle inside the gnomon's arm. Pygonopolis had already pointed out the squares on the other two sides, one in the corner of the original gnomon, the other embraced by it.

One could easily see the three squares that entered the relationship, but what was the relationship? In the new diagram, the area of the large, all-enclosing square was divided into two parts—namely, the tilted square and, just outside it, four copies of the right-angle triangle. If you subtracted away the four copies, only the tilted square remained. In the previous diagram, you could subtract away the two gnomon arms, each amounting to two of the right-angle triangles. In other words, the two diagrams dissected the original square in two different ways. It followed that if you subtracted the equivalent of four of the right-angle triangles from each, you would be left with exactly the same area. The remaining area of the first figure was simply the sum of two other areas—namely, of the squares on two sides of the right-angle triangle. The area remaining in the second figure was simply that of the tilted square, the square of the hypotenuse of the right-angle triangle.

"It's beautiful!" I blurted out. "I never fully realized before . . ."

"You now understand why it was that Pythagoras sacrificed immediately another ox."

"I never realized how much was known about the discovery," I said. "Has this been published anywhere?"

"Ah," sighed Pygonopolis, "not exactly. The proof I just sketched is the one traditionally imputed to Pythagoras and it probably is his proof. But the sequence of events leading up to it, the exploration of the atomogons, then the long pause for thought before the key insight, these are my own speculations. The long pause, for example, must surely have been there. It would take anyone a while to realize that the large square, the one he dissected in two different ways, was not destined to play a role as one of the squares entering the relationship, as it did in his exploration of the atomagons. All told, I simply think this was the most natural sequence of events in the chain of insights that led to the Pythagorean theorem." He seemed sad, as though haunted by a memory.

"Is anything wrong?"

"At the age of 53, Pythagoras ran afoul of the local ruler, Polycrates. The issue is shrouded in mystery, but I would guess that Pythagoras had already begun the quasi-religious activity we know today as the Pythagorean School. Perhaps Polycrates felt threatened by Pythagoras's increasing influence. In any event, Pythagoras immigrated to Crotona, a Greek colony in the South of Italy. There he founded the school that was more than just a school. The Pythagorean School, a secret order that included women, taught that number was the basis of reality and swore its members to secrecy regarding any discoveries, old and new. For example, the discovery of irrational or incommensurable numbers was kept a secret until one of the members of the school leaked it. He was punished, it is said, by accidental drowning in a shipwreck.

"More than this, the school had a distinctly spiritual orientation. They wore white and practiced asceticism. They had five-pointed stars tattooed in the palms of their hands. The school eventually became famous for its doctrine of the transmigration of souls, an element that may have been borrowed from the ancient Hindu doctrine. The body was the temple of the soul, and it would be reborn in an animal if it sank to an animal state of self-gratification. But if it achieved perfection, it would escape forever the endless round of reincarnation, residing at last with the gods. Because some animals were therefore reincarnated humans, it was forbidden by the school to eat the flesh of animals."

"The first vegetarians?" I asked.

"Well, they probably didn't eat mammals, anyway. The Pythagoreans also taught the connectedness of all things, that the human mind was connected to the cosmos itself. It prepared them for life among the gods."

The idea of a connection between the mind and the cosmos reminded me of my own questions.

"Assuming that everything you've said is true," I asked, "how does the Pythagorean adventure reflect on the issues we discussed last night? For example, why do we, living in a completely different culture from that of ancient Greece, find the proof just as convincing as Pythagoras did? Second, did Pythagoras create this theorem, or did he discover it? Third, how does the holos come into all of this?"

"Surely we find this proof convincing today because it's true. Our culture may be different, but the actual content of the Pythagorean theorem is what I would call *transcultural*—that is, beyond or above culture. I realize that you feel compelled to ask such questions, but you must remember my wheel analogy, a good example of transcultural content. The wheel does what it does, more or less equally well, in all cultures that have it.

"If I were to be very strict about it, I would say that the use of dots to represent numbers, even the use of the Greek language to express the proof, are both cultural elements, but the same thing is true today. Look in any mathematics journal. Proofs may be written in English or Chinese, the notation may differ somewhat from author to author, but everyone knows what is meant by the words and notation. In short, there is a kind of translation process lurking in the background, whereby the work of two authors who happen to find the same theorem may be converted, one into the other, almost made interchangeable by the translation process."

"Cultural influence aside," I pursued, "did Pythagoras discover his theorem, or did he create it?"

"He discovered it, of course. Rest assured, if Pythagoras had not discovered the famous theorem named after him, someone else would have. In this sense, the theorem was preexistent. My goodness! What else can you say? If Christopher Columbus had not sailed west in 1492, someone else would have in 1496 or some later time. Indeed, the Vikings had already discovered North America long before Columbus. North America preexisted in a sense that is essentially no different from the preexistence of the Pythagorean theorem.

"As for creation, what is the probability that two painters who do not know each other will both paint the Mona Lisa? Zero, my friend, zero. *That* is creation! And talk about the influence of culture on mathematics! These days we have the cultural fad of creativity. So there are mathematicians, not yet many, who wish to be seen as creative. Without realizing that they are already creative in finding the paths to theorems, they insist on being something like artists, their theorems being works of art, so to speak."

"That seems harmless enough, doesn't it?" I commented.

"Our present culture has become increasingly uneasy about restrictions, whether in the form of authoritarian views or absolute ideas such as Right and Wrong, with capital letters. And it finds the impersonal quality of mathematics repellent. Why, I don't know. It is important to have something from outside that challenges you. It is important to fail."

Pygonopolis was quite out of breath and somewhat worked up. He sat down, while I stared out the window. Then he began to speak, very quietly this time.

"Let me show you how the Pythagorean theorem affects us today."

He stood up, walked to the chalkboard, and wrote the following formula:

$$X^2 + Y^2 = Z^2$$

"Here is Z, the hypotenuse of a right-angle triangle, and here are X and Y, the other two sides. In any coordinate system, where X and Y are coordinates of a point, this formula enables us to compute the distance between two points. Some computers are doing this all the time. Think of all the navigational computers on aircraft, ships, satellites, not to mention the thousands of land-based computers doing distance calculations at any moment. I dare say that the Pythagorean formula is one of the most widely used formulas in the world today. Here, too, is auxiliary proof that the theorem is true. If it weren't, how many planes do you think would land safely? None!"

Something I had mused about before falling asleep the previous night came back to me: "Speaking of computers," I interposed, "it occurred to me recently that we still live in a Pythagorean world, the first cosmos of integers and rational numbers that Pythagoras dreamed of."

"Yes?" Pygonopolis sat upright.

"In a practical sense," I explained, "we still live in the Pythagorean world of rational numbers. We never actually use irrationals when we're measuring or computing things. For one thing, an irrational number has an infinite number of digits and no computer has an infinite memory. So we are forced to approximate an irrational number such as the square root of 2 by a rational number such as 1.4142, which is close enough for practical purposes. This number is rational, of course, because it's the ratio of two integers, 14,142 and 10,000."

"Imagine!" said Pygonopolis. "The world of Pythagoras lives again in computers. What a wonderful idea! I will use it in my next course on the history of Greek mathematics."

That night was to be my last in Athens. Pygonopolis picked me up at my hotel, and we went out for some traditional Greek food. A good supper always brings out the philosopher in me. When well fed and comfortable, what is there to do but ponder the state of the cosmos—or the holos, as the case may be?

"Do you want to go to the holos with me?" he asked, coyly. "It is very easy. We can go there right now." He had been drinking retsina. Was he about to do something foolish?

"To visit the holos, you must assume some axiom system. We have not talked about axioms, yet all the Greek mathematicians realized, to varying extents, that a system of assumptions or axioms underlay their work. By the time of Euclid, a few hundred years after Pythagoras, the axiomatic method was well established. This meant basing all theorems solidly on axioms or on other theorems that were so based. Euclid drew up a list of axioms or postulates on the basis of which all the theorems in his *Elements of Geometry* can be derived. They concern geometry, of course, but this puzzle embodies another, much simpler set of axioms.

"In any event, once in the holos, once you have adopted a set of axioms, you move about, in a manner of speaking. You will quickly find there are places you can go and places where you cannot go. You bump your head on something harder than stone, something so permanent it has always been there and always will be."

"What exactly do you mean by moving about?" I asked.

"You start with the axioms and then look for truths based on them. If you think of something that may be true, you can try to deduce it from the axioms. Failure to do so may be due to your own limitations, of course, but they may also be due to the holos itself. What you think is true may not be. On the other hand, you may succeed in deducing something from the axioms, something that is therefore true. In such a case, you have moved, in a manner of speaking, from the axioms to new ground.

"It is not very easy to come up with a simple axiom system that demonstrates these principles. In fact, I must resort to a puzzle to demonstrate the idea."

Pygonopolis reached into his suit and withdrew a pen. Taking one of the expensive linen table napkins, he sketched the following map on it:

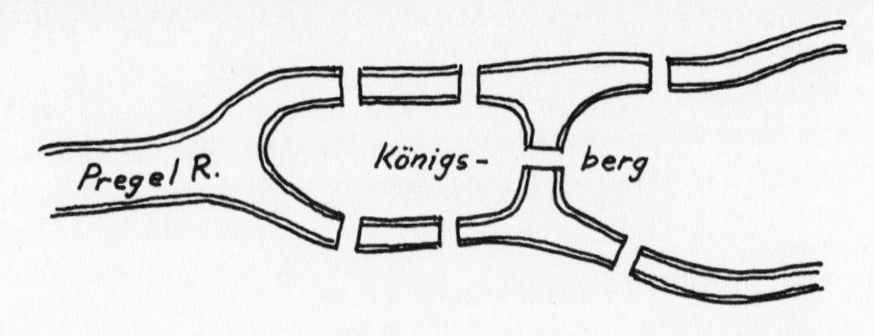

The Bridges of Königsberg

As he sketched, he muttered under his breath. "People laugh at puzzles. If I try to use a puzzle to make a point about mathematics, they think I am joking. But very often, quite serious mathematical systems turn out to be puzzles and vice versa. For example, this pretty little puzzle intrigued the great German mathematician Leonhard Euler. In solving it, he was led to formulate the outlines of a new kind of mathematics, which today we call 'topology.'

"Here we have the ancient town of Königsberg, occupying two banks of the Pregel River, as well as an island and a peninsula in the middle. Seven bridges connected various parts of the city together. It was a recreation of the good citizens of Königsberg, while taking their Sunday strolls, to find a tour that would cross each of the bridges once and only once.

"Before we look at the underlying axiom system, I must visit the lavatory. While I am away, you can try to solve the puzzle."

"What must I do?" I was a bit confused by this game he had put before me.

"Merely pretend that you are a citizen of Königsberg. Start where you like and use the pen to trace your route anywhere on the map but into the water. Trace a route that takes you across each of the bridges once and only once."

"Do I have to end up where I started?" I asked.

"A good question." Pygonopolis shifted uneasily from one foot to the other while he pondered a moment. "In fact, I will let you start where you like and end where you like, but a genuine tour starts and ends at the same place." He shouted as he crossed the restaurant, drawing not a few stares from other diners, "And don't forget, you must cross each bridge once!"

I picked up the napkin with a sigh. I had no pen, so I tried to trace a route with my finger. Although my first few attempts to find a tour that crossed each bridge once failed, I put it down to forgetting where I had already been. After a few minutes, however, I got rather good at remembering my route.

"Well? Are you done?" Pygonopolis had returned.

"Frankly, I've never been much good at puzzles," I said, feeling just a trifle irritated.

"Ha. It doesn't matter how good you are at puzzles. I could stand in that washroom until the entire Aegean went down the drain and you'd never achieve a solution!"

"I suppose you can prove that," I responded increduously.

"I can indeed. But first let me explain the axiom system. It concerns points and lines, almost like geometry, but the lines can be as wiggly as you like, their main role being to connect points."

He turned the napkin over and wrote the following axioms:

1. A network consists of a finite number of points and lines.
2. Every line in a network joins two points.

"These axioms lay the ground rules for something called a 'network.' It consists of points and lines, points being what we call 'primitive elements.' They are not defined but may be interpreted in the normal way. A line, on the other hand, is defined in terms of points. It is something that joins two points. I could be more precise here by saying that a line consists of a pair of points. In any event, these axioms put us at the threshold of a virtual universe of networks. Our job as mathematicians is to explore truths about networks. In the process, we might discover interesting structures for which we invent special names. We make definitions, as I shall do in a moment."

I stared at the axioms then turned the napkin over. I could see little connection between the axioms and the map, except that Pygonopolis had used lines to draw the map. Further, I could see no points to speak of.

"I'm sorry," I said. "I don't quite see what the axioms have to do with the map of Königsberg."

"That was the genius of Euler. Look at this."

Pygonopolis seized the napkin and drew a point in the middle of each of the four landmasses. Then, for each bridge, he drew a line that crossed the bridge from one landmass to the other, connecting their respective points.

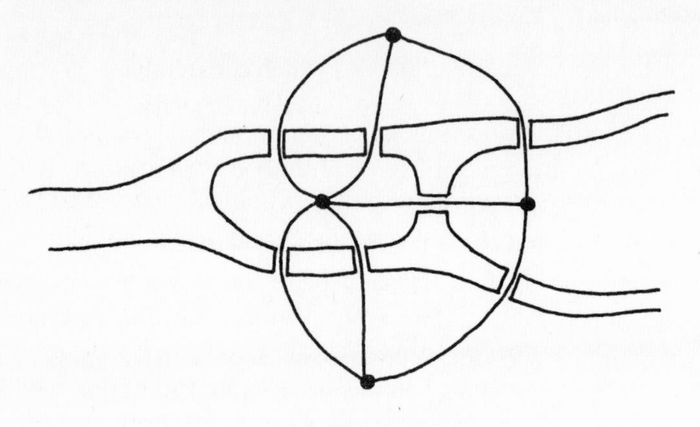

. . . with Network Superimposed

"See how this little figure cuts to the bone of the problem. Any tour you might like to make of Königsberg can be reduced to an equivalent tour of the Euler network. You can meander all you like in a real tour, but once on a particular piece of land, you are effectively at the point that represents it. Every trip you might make with your pen can be reduced to a sequence of lines on this network.

"Now it was immediately plain to Euler, once he had drawn this little network, why the good citizens of Königsberg had such a difficult time touring the seven bridges. See here: If I am touring the town and I come to a point along one line, I must leave it along another. Indeed, in a proper tour, I must use an even number of lines at the other points. And see here. All four points have an odd number of lines connected to them. Ergo, no tour exists. Ergo, the puzzle has no solution.

"Now, as you are already aware, this particular diagram of Euler's is an example of a network. It satisfies the axioms of a network, as you may readily check. It consists of points and lines, and each line connects two points. As a mathematician, you might want to explore the holos by asking questions about networks, then seeking answers in the form of theorems.

"Let's say, for example, that you think that every network has a tour. First, you define what you mean by a tour." Pygonopolis flipped the napkin over again and wrote,

> A tour is a sequence of lines in which consecutive lines share a point and every line of the network appears once in the sequence.

"If, as a mathematician, you think you're onto a general truth, you make a conjecture, a kind of mathematical hypothesis. Thus,

> Conjecture: Every network has a tour.

"You may try to prove your conjecture, making of it a theorem in the process, but you may fail, as you certainly will in this case. You may think of a counterexample—such as Euler's little diagram. This, you will recall, is what happened to Pythagoras. He thought that all lengths were commensurable until Thales showed him the Egyptian figure that turned out to be a counterexample. In the present case, Euler's network serves precisely the same purpose. The conjecture is not true for every network because it is not true of Euler's network. The holos has spoken.

"But then you get another idea. You notice that the trouble with finding a tour in the Euler network was due to points that meet an odd number of lines, so you wonder whether the conjecture might be true if you limit it to networks in which all points meet an even number of lines:

> Conjecture: If all points in a network meet an even number of lines, the network has a tour.

"That should certainly do it," I said. "After all, every time you enter a point in such a network, you are guaranteed that you can leave it."

"You could try such an approach," he responded, "casting it in a formal argument, but you would fail again. Here is a counterexample." Pygonopolis drew a new network in which every point met an even number of lines, but it obviously had no tour.

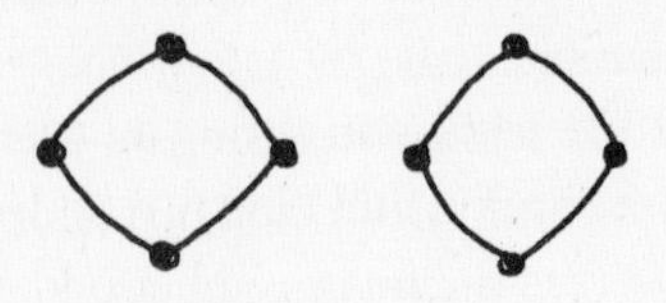

A Counterexample

He had me interested now. I suggested that if the network were connected, the conjecture might work. So Pygonopolis wrote anew on the napkin:

> Conjecture: If all points in a connected network meet an even number of lines, the network has a tour.

"We must of course say what we mean by 'connected.' I propose the following definition:

> Definition: A network N is connected if, for any subset of its points and lines not amounting to the whole of N, there is a line of N not in the subset that meets some point of the subset.

This definition seemed to satisfy our intuitive notions of what being connected ought to mean, but would it be adequate to the task?

"I think I have a proof," cried Pygonopolis. He opened the napkin up to virgin territory and wrote,

> Theorem: If all points in a connected network meet an even number of lines, the network has a tour.

> Proof: One may always construct a tour of such a network as follows: Start at any point of the network and select any line that meets it. This line forms the first member of a sequence. In general, each time a

new line is added to the sequence (including the first one), two cases arise:

1. The new point is not the start of the tour: In this case, only an odd number of lines meeting that point can belong to the sequence, because (a) for every line entering the point, the sequence includes a line leaving it, giving an even number of such lines, and (b) the line just added to the sequence makes the total number odd. On the other hand, the point meets an even number of lines in the network altogether, so there is at least one line remaining, which can be added to the sequence. Let the line be so added.
2. The new point is the first point of the tour: In this case, either the tour is complete or there are lines not yet included. In the latter case, because the network is connected, there must be a line not yet in the sequence but that meets one of the points in it. After all, the sequence as so far constructed contains a subset of the network's points and lines.

Construct a new sequence starting at this point and proceeding in the same manner. By the foregoing arguments, the new sequence must ultimately return to its starting point, because all points of the network still meet an even number of lines that are not in the first sequence. Insert the new sequence into the previous one at the point in question, making a new, larger sequence. If the newly enlarged sequence is not a tour, there will be a line meeting a point of the new sequence that is not in the sequence. Continue to adjoin sequences in this manner until no lines remain to be added. The resulting sequence must then be a tour.

Pygonopolis glanced at his watch.

"By the gods! I have kept you a very long time. You have been very patient with me. After all, I know that much of this is old hat, as you might say."

Our trip to the holos was over. Pygonopolis signaled the waiter. "My friend, I have told you what I think about early Greek mathematics, about Pythagoras, and about the holos. I do not know whether I have answered your questions as well as you would have liked, but I have tried. My final word on the holos is this: You don't have to believe that it really exists in some manner, but you will never encounter a counterexample to such a belief."

Pygonopolis left the table to pay the bill and get his car. I glanced at the rumpled napkin lying on the table, ruined by penned-in diagrams. I picked it up and waved it before the waiter. He looked dismayed. "Here," I said sheepishly to the waiter, and handed him 2,000 drachmas, which I hoped would be adequate compensation.

As a souvenir, that napkin was well worth the money.

PART II

THE SUPERIOR WORLD

CHAPTER 3

Al Jabr

Aqaba, Jordan, June 24, 1995

The next day in my grueling schedule, I flew from Athens to Amman, my head swimming with thoughts of the holos, geometry, and numbers. In Jordan, I changed planes for Aqaba, at the head of the Red Sea. Arriving at the tiny airport there, I met my host just outside the customs office. He was Jusuf al-Flayli, an Egyptian astronomer who kept a summer home in the hills above the port. Al-Flayli had made a study of early Islamic astronomy and mathematics. He was a thin and nervous man, given to frequent quoting and a quietly intense manner.

"Welcome to Aqaba, Professor Dewdney. I am Professor al-Flayli. I hope you will call me Jusuf. My son Ahmed would take your bags, but he has parked the car and is nowhere to be seen."

We stood by the terminal building, looking for his son, when we were engulfed by a cloud of dust kicked up by airplane propellers. Someone seized my bags, shouting in my ear. "Yes, Sayed. Keep going. Soon we are there." When the dust cleared, I saw that a boy had my bags. He grinned earnestly at me.

"Let him carry," said al-Flayli. "It is not far, and he needs the money. Allah favors the compassionate." This was evidently not Ahmed.

We returned to the terminal, where we found al-Flayli's son in the small lobby, inspecting brochures. Barely in his teens, Ahmed seemed at first to be a carbon copy of his father, but then he smiled brilliantly when introduced. We followed Ahmed to the airport road, where we found a Land Rover wedged between two other cars. We could load it only after Ahmed had repeated no less than 20 moves back and forth to extricate the vehicle from its prison. I would have thought it impossible. The bag boy ran off shouting, my dinars clutched in his hand. Al-Flayli's face lit up briefly as we climbed into the Land Rover.

I sat beside Ahmed, who drove. We followed the airport road to a highway then turned toward Aqaba. I had looked forward to seeing the famous port, but the road skirted only its northern edge, passing numerous apartment buildings and then an extensive plantation of date palms. We turned onto another highway, which led north. Civilization fell away to isolated houses—low, white, and windowless—then desert scrub. The road gradually ascended the hills behind Aqaba. Al-Flayli sat behind us, pointing to a hill on the right. His house was up there.

Ahmed took my bags, and al-Flayli led me through the house, about the size and shape of a suburban bungalow. A patio balcony opened out toward the south. The setting sun graced the western hills in gold, barely blessing the tops of minarets in Aqaba. The sea was aquamarine, its shoreline a vast parabola with the port at the apex below us.

"Please. You must be tired," said al-Flayli, gesturing toward a couch on the patio. As soon as he said this, I felt a wave of fatigue pass over me, and I sat down.

Ahmed brought over an elaborately carved wooden table inlaid with ivory. It bore several bowls of fruit and a platter of sliced pita, olives, apricot jam, and hummus.

"If you are not too tired, I hope you will favor us with an account of your travels and of your quest for the meaning of mathematics."

Then, in an undertone to Ahmed, he whispered, "Fetch your mother."

Sensing a certain drama in the way al-Flayli had set things up, I began as dramatically as I could. "I have just come from the ancient city of Miletus on the Aegean Sea. There, I met Professor Pygonopolis, who showed me, in one afternoon, the roots of Greek mathematics, both in their culture and in something beyond culture. . . ."

Al-Flayli's wife, Amina, settled into one of the chairs. At every sentence I spoke, she would smile and nod encouragingly. I told them the story of Pythagoras and his initial belief that the universe was governed by integers and their ratios. I told them about the disastrous figure from Egypt (at which al-Flayli smiled), then about the Pythagorean theorem. Al-Flayli was rapt throughout the tale. His son Ahmed kept looking from me to his father and back again, his eyes wide in wonderment. I finished with the holos.

"Frankly, I have never heard that word before," said al-Flayli. "Is it something your Greek friend made up?"

Somewhat taken aback by al-Flayli's insight, I had to admit it was.

"Well, he may have something, in any case. After all, one does need names for things!"

The sun was sinking below the horizon to the west and night poured into the valleys about Aqaba like a rising tide of darkness. The air grew chill. The al-Flaylis excused themselves for the evening prayer while I sat in the small but lavishly furnished living room, amusing myself by looking at the titles of books in their shelves. Many were in Arabic, but many were also in English, including what seemed to be translations, with titles such as *The Walled Garden of Truth, The Parliament of Birds, Majnun, Laila,* and so on.

Soon after prayers, we had a large supper with many delicacies in as many bowls and plates. I ate until I could eat no more, and still Amina pressed more on me. At last, we staggered to the living room to recline on soft cushions.

"So you must now elaborate on your letters and e-mail messages. Tell me, please, how you think we can help you in your quest." He spoke very quietly, almost gently, but there was something of the steel trap in his manner, so I spoke carefully.

"I wonder what you can tell me about the Arab contribution to mathematics, how it was influenced by culture and to what extent. Yet I also want to hear your opinion on the noncultural or transcultural element, as Professor Pygonopolis called it, whatever it is in mathematics that could be said to have originated outside us. Is mathematics invented or discovered, or is it some combination of the two? And, to the extent that it's discovered, does this explain what some have called its 'unreasonable effectiveness' as a description of the world? Let me be clear, if mathematics has a prior existence, can it —"

"Very well. Now I have it. There you are, creeping toward what you call the holos, but too shy to mention it, like someone who doesn't dare to believe in a thing that is too good to be true. Well, I suppose you, indeed all of us, have a right to be suspicious of this state of affairs. But let me address these questions as they crop up in the little story I can tell about the development of mathematics in the Middle East from the tenth century of the Christian calendar to the fifteenth and sixteenth centuries. Indeed, I must remind you that the Islamic empire flourished and persisted from the seventh to the sixteenth centuries, holding sway for more than twice the time the Christian West has dominated, from the Renaissance to the present day. In this expanse of time, this 1,200 years, there was a great ingathering of scientific knowledge, both mathematical and physical. In my own profession of astronomy, there was great progress, perhaps the greatest, but all our achievements were crowned by an error, a mistake of perception inherited from the Greeks but common, in fact, to all people of the Earth."

With this, Ahmed turned down the light and al-Flayli put on a tape of Arab music. The sound of oud and ney circled his words, evoking an atmosphere of long ago and far away.

"The story of mathematics in this part of the world has its roots in Babylon, in India, and in Egypt, both under the Pharaohs and later under the Ptolemys. But it first flowered in the intellectual gardens of Baghdad in the early days of the Islamic Empire. That was in the ninth century, during the reign of the Caliph al-Ma'mun.

"In less than two centuries, from A.D. 620 to the year A.D. 800,

Islam had grown from a revelation in Mecca to a religious community that stretched from Spain in the west to Persia in the east. In the process, it had become, de facto, an empire. It had ministers, diplomats, and an extensive civil service. Peace within its borders and wealth within its lands gave rulers like al-Ma'mun creative opportunities. Al-Ma'mun was in some ways a stiff and intolerant man, but he valued learning and wisdom above all else. He sent out scouts to find the most eminent and learned, whether within his borders or without, to join him at the court he called the 'Bayt al Hiqma' or House of Wisdom.

"It was nothing less than a school of thought, a freewheeling university. The invited scholars were given perpetual grants, in effect, and they had access to every facility that al-Ma'mun could provide. He patronized the translation into Arabic of Greek works such as Euclid's *Elements of Geometry,* Archimedes's *Sand Reckoner,* Plato's *Republic,* and Ptolemy's *Almagest*—hundreds of works altogether. There were the Siddhantas or 'collections' of the Indian scientists Brahmagupta and Aryabhata to translate, as well. The new books were avidly read, copied, and distributed, with more appearing all the time. Al-Ma'mun built a great library, where he collected every document from every corner of the Islamic lands and beyond. He also financed the construction of two major observatories and several smaller ones. It was a golden age that outlived the caliph. And in it appeared the greatest Arab mathematician of them all, al-Khwarizmi.

"In fact he was not an Arab, strictly speaking, but Persian, from the town of Khwarazm. Born Mohammad ibn Musa, he went to Baghdad as a young scholar who had already familiarized himself with many number systems then in use around the world, particularly India."

"Most of them began I, II, III, like the Roman alphabet. Here the first three whole numbers are rendered simply as tally marks. Following that are repeated marks of different kinds, usually with separate symbols for 10 and 100 or, in the Babylonian system, for 60 and 600. If you wanted to add two numbers in these systems, it was simpler to count on your fingers and very easy to be wrong. Commerce was a nightmare, relatively speaking.

"Imagine al-Khwarizmi's delight at the system from India. It had 9 separate symbols for the first 9 numbers and a very important new number: 0 [zero]. Moreover, the numbers would repeat themselves in tens in the most pleasing way. More importantly, the numbers lent themselves readily to arithmetic manipulations. The secret lay in the new positional notation, a way of using the positions of digits in a number to enhance their expressive power.

"In this system, every number consisted of digits, and each digit expressed a multiple of some power of 10, as you know. It's so easy to take this number system for granted, for we Arabs have lived with it for over 1,000 years. Yet it is really almost magical, when you think about it. The number 375 actually tells us its own composition if we look not only at the digits, but also at the positions they occupy. It consists of 5 units or 5, 7 tens or 70, and 3 hundreds or 300. It is the sum of these separate numbers. Now a number that can be written as a sum of such parts lends itself to addition and other operations because they can be carried out on one part at a time. To add 375 to 193, for example, you merely added the units together: 5 plus 3 equals 8, so 8 is the first digit; Now add the next two units together. 7 plus 9 equals 16—that is, 6 tens and 1 hundred, so 6 is the second digit; the 1 is carried to the next position, where we must add 3 and 1. This makes 4, added to the 1 carried makes 5, the third and last digit.

"To work in this system, you only needed to know the addition tables for the first 10 digits, 0 to 9. The same thing is true for multiplication. The positional notation would not work without 0, by the way, for without it, numbers would collapse. We might write 704 as 74, and all would be chaos. The new 0 or *sifr*, as it was known in Arabic, puzzled many who encountered the system for the first time. What was the point of a number for nothing? If nothing was there, no number was necessary to count it. Wags of the day had their way.

"The contrast between the new notation and the old suggests to me an almost trivial yet profound observation that bears on both your question concerning cultural influence and the question of whether mathematics is created or discovered. For example, al-Khwarizmi, along with everyone else who used the new number system, knew of other systems as well. The same numbers inhabited all systems, in a manner of speaking. The Roman XLII was an earlier way of writing the number represented in the new system as 42, for example. There was a superficial difference between the two numbers, but an underlying identity. The difference was cultural and invented, but the similarity was beyond culture. I would maintain that it was discovered. What else can you say?"

I felt that in the similarity lay the answer to my question about discovery, so I asked the most direct question imaginable: "Tell me, please, what is the similarity? Can I understand directly and without intervening symbols, the number 42?"

Al-Flayli looked at me and smiled sadly. "Of course I know what you mean, but you are really being too ambitious. Try this with the number 2 or, better yet, 1. And think of the English words *one* and *two.* Or the Arabic *wahid* and *ethnain,* which mean not just approximately but precisely the same things. Or imagine that you are a Roman and say *unum* and *duo*. Can you directly conceive even of these small numbers? I'm not sure you can. You may fool yourself and think you are perceiving the number 2 in its pure form when, in reality, you are imagining 2 dots, side by side.

"Can you offer an explanation of why this is so?" I asked.

"No, I cannot. I can only say that we apprehend numbers only through our notation, our words. But we cannot dispense with these vehicles any more than we can walk without feet. You see, pure number belongs to what some Arab mathematicians called the Superior World. Let me read you something." Al-Flayli reached behind him and, without actually looking, withdrew a book from the shelf behind him.

"This is a translation of a very ancient book called *The Epistles,* a collection of essays on the arts and sciences. It was authored anonymously by members of a school called 'The Brethren of Purity,' who flourished in the tenth century, although there is evidence that they were active throughout the Islamic era." Al-Flayli read,

THE MEANING OF NUMBER

> The form of the numbers in the soul corresponds to the form of beings in matter (or the hyle). It [number] is a sample from the Superior World, and through knowledge of it the disciple is led to the other mathematical sciences, and to physics and metaphysics. The science of number is the root of the sciences, the foundation of wisdom, the source of knowledge, and the pillar of meaning. It is the first elixir and the great alchemy. . . .

"That would be as clear a statement as you will ever get about the underlying numbers. In this view, the numbers exist in the soul, or mind, yet their origin is from outside the mind. Not only does number reside, in a sense, in material objects, but pure number, unattached to any particular thing, arises in the so-called Superior World.

"You must not forget that these scholars were all Muslim and would locate their philosophy within the revelations of the Koran. In other words, the Superior World, insofar as it relates to the truth of things, is nothing less than an aspect of God. Allah has 99 other names, including Al-Haq or the Truth. Thus, numbers and all the truths pertaining to them belong to the Truth of God or Al Haq."

"Is it a place?" I asked.

"Is what a place?"

"The Superior World. Where is it?"

Al-Flayli laughed quietly, a sort of easy chuckle. "Well, I dare say it is very close to the holos, if you happen to know where that is."

"Who then, were the Brethren of Purity?" I asked, pursuing a different angle.

"Scholars, as I said, but scholars with roots in antiquity. In fact, they originated in Mesopotamia but may just represent a continuous chain back to the Pythagorean Brotherhood. Here is what one of them writes:

> Know, oh brother (may God assist thee and us by His spirit), that Pythagoras was a unique sage, a Harranaean who had a great interest in the study of the science of numbers and their origin, and discussed in great detail their properties, classification and order. He

> used to say: "The knowledge of numbers, and of their origin from unity, is the knowledge of the unity of God, may He be Exalted; and the knowledge of the properties of numbers, their classification and order, is the knowledge of the beings created by the Exalted Creator. The science of numbers is centered in the soul, and needs little contemplation and little recollection before it becomes clear and is known without evidence."

Al-Flayli continued, "Well, for sure, Pythagoras predated the coming of Islam, yet we Muslims believe that Islam has always been here on Earth. So the Brethren of Purity would probably identify Zeus with Allah and claim that the gods were actually aspects of deity or emanations of Zeus. However that may be, we find strong hints that the Pythagorean Brotherhood survived well into the Islamic era, little changed in its mathematical and philosophical orientation from what Pythagoras himself taught."

"Getting back to the underlying reality of number, I can only say that there was a kind of operational identity. Proof of this was simple: If a Roman purchased XLII sheep from an Arab shepherd, he was well satisfied when the Arab delivered up 42 sheep, for this was precisely the number the Roman had ordered, neither more nor less. This humble example illustrates the reality of numbers in the world. At the same time, the concept of number in the soul of the Roman and of the shepherd were the same. And by virtue of this, either could recognize 42 or XLII in any other collection of objects, be they stones, fruit, or anything else. If you want to argue about creation and discovery, I would go only this far: The human mind creates numbers in the same sense that it creates colors. Yet the colors we perceive correspond to something real outside of the mind. In this latter sense, we are discovering numbers all the time. How many pages in this book? How many people are on that bus? How many dinars in my pocket?

"Speaking of dinars, we may go one step beyond numbers to arithmetic and examine the new number system brought by al-Khwarizmi to al-Ma'mun's court. When those who learned the new system of arithmetic tried it on money, the advantages were immediately apparent. The sum of two amounts of money appeared, almost as if by magic, under the new addition operation and invariably matched the total found by tally. Money paid out of an account could be readily subtracted from the books. Simple commercial forecasting became possible with the new math. The potential returns of commercial ventures could be readily calculated by the multiplication of goods by price and then the subtraction of expenses.

"A great barrier had been breached. Numbers took on a fluid character, and it became possible to think about them in new ways."

Amina excused herself. "I will see you at breakfast tomorrow before you three depart for Wadi Rum."

Ahmed, meanwhile, had been pressing farther and farther forward on the couch and now was about to fall off. "Papa, tell our friend that story about the House of Wisdom."

Al-Flayli smiled, paused, then said in his quiet voice, "I told this to Ahmed once, and he has never since ceased begging me to repeat it. With your permission, perhaps the time has come to tell it again.

"The House of Wisdom was a special court maintained by al-Ma'mun. We do not know its shape or size, but we imagine a great hall with sand tables for reckoning, a scriptorium, astrolabes, armillary spheres, and other mathematical and scientific instruments on shelves, and a special place where speakers could gain the ear of their colleagues. Here sat al-Ma'mun, clothed in brilliant colors and jewels, applauding the boldest concepts, encouraging his scientists.

"There in the court was Hunain ibn Ishaq, a Christian scholar and physician who translated works from Greek. And there were the Banu Musa, or sons of Shakr ibn Musa, able geometers who collected and translated dozens of Greek manuscripts. There was al-Hallaj, famous for translating Euclid's *Elements* into Arabic. Habash al-Hasib composed extensive tables of accurate astronomical observations and advanced the science of trigonometry. Thabit ibn Qurra, the Astronomer Royal, ran the observatory in Baghdad and had made numerous mathematical discoveries. Al-Kindi and al-Farghani wrote the first extensive treatises on astronomy at this time. Al-Nairizi wrote a commentary on Ptolomy's *Almagest* and developed the spherical astrolabe. There were also many poets and artists, including al-Mawsili and his son, musicians to the House of Wisdom.

"What was it like, this House of Wisdom? In session, he who had the floor might expound the theory of harmony in vibrating strings, then demonstrate octaves, fifths, and thirds on the oud. The musicians would begin and end on these chords, elaborating cleverly convoluted passages of melody between. All those present would be rapt in the blend of intellectual and aesthetic beauty.

"Then another would step forth: 'O Commander of the Faithful, Shadow of God's Will on Earth, Light of the Eye: I would present to your eminence and to this distinguished court one newly arrived from Khwarazm in Lower Mesopotamia. His name is Mohammad ibn Musa al-Khwarizmi, and he hath somewhat to tell us of number and of the systems by which one may make them.'

"Al-Khwarizmi's presentation of numbers and their systems stunned the caliph. Other members of the House of Wisdom, most able mathematicians, grasped the new idea at once. Al-Khwarizmi became the new favorite of al-Ma'mun. Later, perhaps within a year, Al-Khwarizmi completed a wonderful book and dedicated it to the ruler. I will not torture you with the Arabic, but in English the title would be *The Comprehensive Book of Calculation by Balance and Opposition*. In particular, though, you will want to know just a little Arabic. The word *balance* is a translation of the Arabic *al jabr*, the word that has become today's *algebra*.

"The basic nature of algebra can be found in the equation. There is an equals sign, real or implicit, that joins two expressions. The expressions may look different, or be described differently, but their relationship, the equality, produces powerful restrictions. Now the word *opposition* in the title refers to the two expressions that appear so different. Yet the word *balance* refers to the equality of the expressions. The balance is maintained only by treating both expressions exactly alike. What you do to one, you must do to the other. If you subtract a certain amount from one expression, or multiply it by a certain amount, you must do exactly the same to the other expression. Thus, if the expressions were equal before any of these operations, they would be equal afterward."

Al-Flayli produced a paper and pen, writing on it the following equation:

$$(1/12)\,X^2 = X + 24$$

"This is an example from al-Khwarizmi's book, written in modern notation. Al-Khwarizmi solves the equation by first multiplying both sides of the equation by 12, producing

$$12 \times (1/12)X^2 = 12 \times X + 12 \times 24,$$

which becomes

$$X^2 = 12X + 288.$$

"Then he subtracts $12X$ from both sides of the equation to produce

$$X^2 - 12X = 12X - 12X + 288,$$

which equals

$$X^2 - 12X = 288$$

"Then he notices that if he adds 36 to both sides, he will obtain something very interesting:

$$X^2 - 12X + 36 = 288 + 36 = 324$$

"Now the expression on the left is a perfect square, namely $X - 6$ multiplied by itself. That is, if you multiply $X - 6$ by $X - 6$, you get $X^2 - 12X + 36$. The expression on the right-hand side, a simple number, is also a perfect square, namely 18^2. If two squares are equal, so are their square roots, so

$$X - 6 = 18$$

"When you add 6 to both sides, you finally get the solution:

$$X = 24$$

"I realize that all of this is a bit tedious, yet in every step we have observed the principle of balance or equality and then, almost by magic, the solution appears. There is one and only one number that satisfies the equation—namely, 24. At the beginning, al-Khwarizmi does not know what that number might be but calls it *X*, the simplest but mightiest invention of mathematics.

"Now I do not want to mislead you, so prepare yourself for a shock. Al-Khwarizmi did not use *X*, nor did he use equations. Everything was in words. Instead of *X* he used the Arabic word *shay,* or thing. And in setting out the problem, he would say something like this:

> One-third of the thing multiplied by ¼ of the thing produces the thing with 12 added. This makes 1/12 of the square of the thing, and so the square of the thing equals 12 times the thing, with 288 added.

"I won't bore you with the whole transcription, but that gives the flavor of his work. What do you know? We have here another example of culture in mathematics. You will see two quite different mathematical creations. One consists of symbols, the other of words. Now it is quite an easy matter to translate one form into the other. Someone who did not recognize this translatability could overexaggerate the importance of the difference, but the difference is superficial. Under the surface are the same ideas expressing the same restrictions on this unknown thing that we call *X*.

"The wonderful thing about *X* is the act of faith we make when we say, 'Let the answer be called *X*,' as though conjuring it forth from the Superior World. But what he calls forth is not of his own choosing. He must accept what appears. It is the art of the magus, or ancient magician."

Al-Flayli lapsed into a silence and his eyes floated toward the ceiling. I made bold to interrupt the meditation. "Considering the real convenience of modern notation," I asked, "why didn't al-Khwarizmi or one of his colleagues stumble on the possibilities of symbolism?"

"I have wondered about that myself. I suppose it was culture that prevented us from abandoning our language. As one wag of twelfth-century Damascus put it, 'The nations of humanity have three excellences: the brain of the Frank, the hand of the Chinese, and the tongue of the Arab.' There you have it. In the House of Wisdom and in other venues as well, some of us could not refrain from expressing scientific ideas and poetry in the same breath. How could we do that with the purely symbolic *X*, not to mention symbols for addition, squaring, and cubing?

"I will say this. For a thousand years, we were collectors, custodians, and improvers of mathematical knowledge. We made many contributions of a practical kind and a few theoretical or general things. Omar Khayyam, for example, completely solved the general equation of third degree. In short, we were well aware of the process of generalization, but we were also in awe of this thing before us. It was more than simply a manipulation of symbols or words, it was a form of contact with something far beyond us, something both adamantine and ephemeral. In it we could smell the perfume of the Superior World."

Up to this point, young Ahmed had been quietly patient, but finally he could restrain himself no longer. "Baba, you did not finish the story of the House of Wisdom."

"That is true, Ahmed. I became much too involved in the happiness of our guest." (He narrowed his eyes at Ahmed as though chiding him, then abruptly smiled.) "It is time for you to go to bed. But I will finish the story tomorrow night when we are all together out on the desert."

Ahmed went off to bed, and I sensed that my time with al-Flayli that evening was rapidly diminishing. There was just enough time to tease out the true influence of culture on Islamic mathematics.

"Are you telling me," I asked, "that the early Arab mathematicians could solve tricky mathematical problems pretty much as we do today without being affected—I mean without the mathematics itself being affected—by metaphysics?"

"If by metaphysics you mean the philosophy of the Brethren of Purity, I would say yes. There was a clean separation between mathematics and its philosophy because the early mathematicians recognized from the start that deduction and nothing but deduction was to be applied to whatever definitions and axioms were given one. However, there were opinions, widely accepted opinions, about what we could call the personality of numbers."

My eyebrows shot up, and al-Flayli smiled at the sight.

"Is it possible you didn't know that numbers have personalities? They stood for things besides mere quantity. For example, the number 1 stood for unity, from which all the other numbers proceed. As such, the number 1 stands for Allah, Who is One. The vertical stroke of the 1 is almost the same as the Arabic *aliph*, the first letter of the alphabet, also a vertical stroke, and the first letter in the name of God. The number 2, the first even number, stood for duality, or the creation. The number 3, which was visually symbolized by a triangle of dots, represented harmony, while the number 4, a square, stood for stability. And on it went for quite a while.

"Indeed, have you heard of amicable numbers?" A bell rang somewhere in my head. Was it from a course in number theory I had taken as an undergraduate? The definition did not come fully to mind.

"Two numbers are said to be amicable with respect to each other," al-Flayli continued, "if each is the sum of the other's factors. For example, 220 and 284 are amicable. The divisors of 220 are 1, 2, 4, 5, 10, 11, 20, 22, 44, 55, and 110. These add up to 284. On the other hand, the divisors of 284 are 1, 2, 4, 71, and 142. These add up to 220.

"To be honest with you, I have no idea what role amicability played outside of this definition. I could guess that someone who had lost a friend might wear an amulet with 220 and 284 engraved upon it to restore the lost friendship. That sort of numerical magic was fairly common in the ancient world, and not just in Arabia.

"But I do know that amicable numbers fascinated many mathematicians of the time. I mean, quite apart from magic, such pairs of numbers posed real challenges to more modern mathematicians who knew nothing of the magical connection. Indeed, great European mathematicians such as Fermat, Descartes, and Euler studied the problem of finding all pairs of amicable numbers. They could barely improve the method discovered during that first golden age of Islamic mathematics.

"In the days of the House of Wisdom, it was Tabit ibn Qurra

who made extraordinary progress with this problem. Here is his theorem." Al-Flayli produced a paper from a historical journal.

> Theorem: If the number p has the form $3 \cdot 2^{n-1} - 1$, q has the form $3 \cdot 2^{n} - 1$, and r has the form $9 \cdot 2^{2n-1} - 1$ and if all three numbers are prime, then the numbers
>
> $$m = 2n \cdot p \cdot q \quad \text{and} \quad n = 2n \cdot r$$
>
> are amicable.

"You may use ibn Qurra's formula to generate many pairs of amicable numbers. However, their size grows very quickly. For $n = 4$, for example, the numbers p, q, and r become 23, 47, and 1,151, respectively. Notice that in this case all three numbers are prime: They cannot be divided evenly by any numbers except 1 and themselves. So the theorem applies. Now if you insert these values for p, q, and r into the formulas for m and n in the theorem, you get m as the product of 16, 23, and 47—namely, 17,296—and you get n as the product of 16 and 1,151, or 18,416." He wrote the two amicable numbers on a sheet of paper for me to contemplate:

17,296 and 18,416 are amicable

"Pierre de Fermat independently rediscovered ibn Qurra's theorem. This was not a coincidence, of course. Mathematical theorems are being independently discovered all the time. This is because, in my humble view, they are waiting to be discovered, perhaps in the Superior World. In any event, after Fermat, Descartes used the theorem to find another pair of amicable numbers:

9,363,584 and 9,437,056

"As far as I know, the amicable numbers go on forever."

I pressed on with al-Flayli. I also wanted to know more about algebra. If the Arabs were so practical, how did they actually use algebra?

"The science of al jabr was especially important in solving a great variety of practical problems such as the division of land, construction work, commercial transactions, and what have you. For example, a man has enough money to purchase 1,760 building blocks from which he hopes to build a house that is twice as long as it is wide. If the walls are to be 8 blocks high, how big a house can he build with those blocks?

"We begin, arrogantly enough, by assuming that we already know the answer. Let the shay or X be the length of the short side of the house, as measured by the blocks themselves. The area of the floor will therefore be $2X^2$ blocks, and the walls will require 8 times the perimeter of the house, which is $6X$, for a total of $48X$ blocks. We therefore have the equation

$$2X^2 + 48X = 1760$$

which we can begin to solve by using al jabr to simplify both sides. In short, we divide both sides of the equation by 2. This yields up

$$X^2 + 24X = 880$$

"We use al jabr again to add -880 to both sides, reducing the equation to

$$X^2 + 24X - 880 = 0$$

"Now, I must say, we have been lucky because, as it happens, we can write the left side of this equation as the product of two factors:

$$(X + 44)(X - 20) = 0$$

"When a product of two numbers is 0, then one of the numbers must be 0. It follows that either $X + 44$ equals 0 or $X - 20$ equals 0. The first possibility leads to X being -44, which is meaningless for the problem at hand. The other possibility leads to X being 20, meaning that the short side of the house must be 20 blocks wide.

"Just to check, al jabr has told the man that the dimensions of his house will be 20 blocks wide, 40 blocks long. Going back to the separate terms of the equation, the floor will use $2X^2$ or 800 blocks and the walls will use $48X$ or 960 blocks. The total number of blocks used will therefore be 800 plus 960, which is 1,760, exactly the number of blocks the man will buy."

"What about the doors and windows?" I asked.

Al-Flayli laughed. "It's true, I left them out, didn't I? Well that can be fixed easily enough. But, truly, you see here the flavor of the method. It is not so different from what high-school students learn today.

"Are you tired?" he asked suddenly. Knowing that he might well want to turn in, I agreed that I was. Then he surprised me.

"If you are not too tired, there is one other thing I would like to show you about algebra, but this relates in a most interesting way to the geometric designs for which we Arabs are famous."

He reached behind himself with that same unerring aim, and retrieved a large-format book of color photographs and reproductions. When he opened it, I couldn't help gasping. There were amazing designs composed of repeating patterns, some geometric, some floral, but all somehow informed by the same crystalline spirit.

Islamic Wall Design

"These designs reach for the infinite in the sense that they go on forever. Even when restricted to a wall or a floor, one can see that the same pattern is capable of limitless reproduction. In this way, everyone who views such designs is invited to consider the infinite, a quality pertaining only to Allah and, perhaps, the Superior World.

"Now from a mathematical point of view, the designs are clearly geometric, yet there is algebra lurking here, as well. When you consider the symmetries of these patterns, you come to the notion of what we know today as a group."

"Don't tell me the Arabs discovered groups! I thought this concept didn't arise until the eighteenth century."

Al-Flayli laughed again, showing little sign of tiring. "No, no. Not at all. Yet they made what you might call an implicit discovery. But before I explain that, we had better review what these groups are."

"Look closely at this design. If you shift the whole thing upward by the width of the repeated pattern that composes it, you get exactly the same design again. This is a symmetry operation. It's called a *translation*. Also, you can see another translation, this one going to the right. If you shift the whole design to the right, it comes into coincidence with itself, just as it did when you shifted it upward. This is another translation. Can you see any other symmetries in this design?"

"I can see that it is also a mirror image of itself," I observed.

"Yes. This is called a *reflection*. You may flip the image 180 degrees, turning it out of the page and back into the page, and find the same design reappear. Now you can do this flipping only about certain lines, called *lines of reflection*, as you can see:

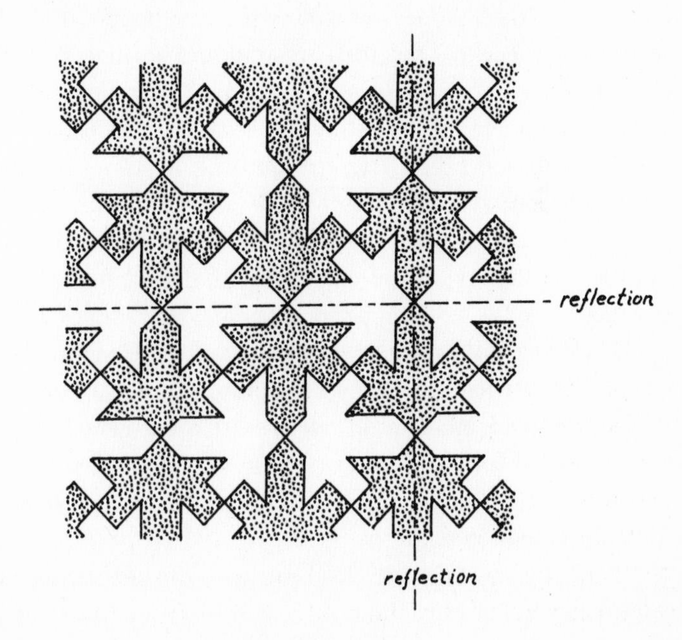

Lines of Symmetry

"There is another operation, called a *rotation*. It's self-explanatory. There are certain points where you can rotate the whole design about by 90 degrees, and you get the same design back again. The point is that these symmetry operations have a very interesting property that I believe would simply not occur to anyone in the ancient world. If you follow one symmetry operation by another, you get a third symmetry operation. In other words, you can treat these operations like symbols, multiplying them together.

"More than this, these operations obey certain mathematical laws." He ticked them off on his fingers:

"One. The product of two symmetry operations is a symmetry operation, as you have just seen.

"Two. There is a null symmetry operation that does not move the design at all. That is clear. Just do nothing. This may seem pointless until you hear the next law.

"Three. For each symmetry operation, there is an inverse operation. Do you know what I mean? In other words, for any symmetry operation, whether a translation or a rotation, there is another one that undoes everything the first operation does. The result of multiplying a symmetry operation by its inverse is the null operation, of course. The result of this particular multiplication of a symmetry operation by its inverse is the same as doing nothing at all. The null operation plays the same role in groups that the zero plays in numbers. In fact, groups generalize numbers.

"There is a fourth law, which says, in effect, that when you carry out three symmetry operations in a row, it doesn't matter how you put them together. You can perform the first operation and follow it by the product of the second two. Or you can perform the product of the first two, then follow it by the third operation. I fear I did not explain that well, but it doesn't really matter. The main point I want to make is just this. These four rules are simply the axioms of what we call a group. Different kinds of symmetries in a design lead to a different kind of group.

"Now, not all designs in this book have the same group, as you might suspect. For example, some can be rotated by 60 degrees, others only by 90 degrees. It was proved in the nineteenth century that the number of possible symmetry groups is finite. There are only so many of them—17, in fact."

"Really? I would have guessed an infinite number," I commented.

"Not at all. Just 17, no more, no less. And every one can be found somewhere or other within the borders of the former Islamic Empire. Every one."

"I see what you mean now." I remarked in surprise. "In other words, the early designers of these marvelous patterns could never achieve a design with basic symmetries that strayed outside of the 17 possible types, yet they actually found all of them?"

"Exactly."

"That raises the question of whether the designers knew there was a limit."

"An excellent point," said al-Flayli. "There were some geniuses among them, undoubtedly. And what a blend of genius: half art, half mathematics! There are broad hints in some designs that the artist was striving to incorporate forbidden symmetries into them, only to be rebuffed by the Superior World, if you like. For example, in some designs you will find fivefold symmetry, little patterns that can be rotated by 72 degrees, five of these rotations making up a full 360-degree rotation.

"Now that particular symmetry cannot occur as a symmetry of any such group. The artist gets away with it by ensuring that all of the real symmetry operations carry one forbidden figure into another. Here is what I mean." He opened the book to a certain page and held it up.

He placed the book on my lap and then stood before me. "I really owe you an apology. I indulged my pride in my heritage instead of considering your comfort, which really ought to come first. Will you forgive me?"

This was perhaps al-Flayli's way of heading for bed, so I stood up, too.

"There's nothing to forgive," I assured him. "How could I be less interested in such things than you are?"

Al-Flayli laughed out loud, an unusual thing. "We shall make an Arab of you yet!"

Fivefold (Fake) Symmetry

He showed me to my room. It was rather cool. The lone window, which looked west, revealed the setting crescent moon and admitted its cold light.

"You might consider," said al-Flayli in the quietest of tones, "watching the moon as it sets. You will see it easily from your bed. Consider that the moon does not shine by its own light, but by that of the sun. The early Arab astronomers knew that. For the moon represented Mohammed, who shone not by his own light, but by that of Another. That is why the crescent moon has such a special place in the symbolism of Islam.

"Tomorrow we go to Wadi Rum," said al-Flayli as he closed the door.

I normally do not sleep well in other people's houses, but here I felt at home for reasons I could not put my finger on. The moon was beautiful, reddening as it sank behind the hills to the west, and I fell soundly asleep.

CHAPTER 4

The Spheres

The morning sun was so brilliant I could scarcely make out the details of the balcony around me. Before I could quite finish a pita slathered with apricot jam and cream cheese, Ali and Ahmed excused themselves to prepare for the trip. I took my coffee to the front of the house to watch them load sleeping bags, thermoses, and extra blankets. At last they were finished and al-Flayli returned to the house.

"I want to show you something before we leave," Ali said. We went into his study, where a peculiar-looking brass instrument stood on a pedestal in one corner. It consisted of numerous circular bands that formed a hollow sphere.

Armillary Sphere

"This is called an *armillary sphere,*" he explained, "an ancient instrument that embodies a great deal of important knowledge about the night sky, the planets, and the stars. It has an equatorial band running around its middle and an ecliptic band set at an angle to it. I will explain these terms tonight, but just for now, I want you to examine its spherical shape. It is an almost literal model of what ancient astronomers supposed the heavens to be like."

"You mean a sphere?" I asked.

"Exactly. The idea that the stars are attached to a huge, rotating sphere is an illusion, of course, but a most important one. In fact, a sphere is a perfectly good model for stellar positions if you're not worried about how far away individual stars may be. Specify their positions on an imaginary sphere with Earth at its center, and you have specified where astronomers, ancient or modern, can point their instruments. For this purpose, the spherical model is perfectly adequate. Today we call this abstract model the celestial sphere. As you can see by examining the equatorial and other bands, there are degrees marked on them. In fact, the position of every star in the night sky can be given in terms of just two angles, like latitude and longitude. This, too, I shall explain later."

The armillary sphere had an unmistakable antique look. "It must be very old," I opined.

"The original," stated al-Flayli in his quiet manner, "may be found in the British Museum. This is an exact copy of an instrument that dates to thirteenth-century Persia."

"It is time to leave," al-Flayli remarked abruptly. He bustled with uncharacteristic energy. We got into the Land Rover, al-Flayli lingering to speak briefly with Amina. She smiled and waved as we left, saying, "I hope you are safe on the camels."

As we drove out the driveway and down the steep access road, I asked al-Flayli what she had meant.

"Only that camels look very nice in the movies, and beginners fancy they would enjoy being on one. But when they get up there and realize how high it is, many people want immediately to get off. A trip of any length simply terrifies them. I wonder if you will be like that." He smiled shyly.

I worried about this possibility every five minutes, on average, for the next hour. In that time, we made our way through the hills and down into a flat expanse of desert scrub dotted with low, waxy-looking bushes. An abrupt turn off the highway, which Ahmed took at a speed nearly sufficient to roll the Land Rover, made me forget all about camels. Now we followed a stony track that was visible only as two faint scars winding into the middle distance, where massive buttes loomed. The buttes were purple and ocher, magenta and brown, all fading into the distance, colors muting to gray. It was astonishing, like a moonscape.

"We are coming soon to Wadi Rum," al-Flayli shouted through the noise and dust. The road rounded the corner of one of the buttes, now towering into the sky on our right. It revealed a vista that I shall never forget. Ahead of us, the *wadi,* or canyon, wound due south, a vast, flat floor of stones and sand presided over by a line of cliffs that marched almost to a vanishing point.

Within half an hour we arrived at a huddle of black goat-hair tents. An old man came running out from one of them. Al-Flayli explained: "He is the sheikh of the Bani Harith, a nomadic tribe that has been living here for several years now."

We descended from the Land Rover and, while al-Flayli and the sheikh discussed details of the trip, Ahmed took me to some ruins, peacefully decaying against the canyon wall behind the settlement.

"Here, you should know, was once a Roman fort. So that is why they call it Wadi Rum, or Wadi of the Romans. Here you can see the remains of a bath and over there was a barracks once standing."

The men of the tribe whom al-Flayli had hired as guides were not overeager to begin. The trip would be short, by their standards, explained al-Flayli, and they disliked being out in the heat as much as anyone. So we sat in the tent of the sheikh, drinking cup after cup of thick black coffee, while the sheikh smoked cigarette after cigarette. The Bedouins listened intently as al-Flayli spoke to them in Arabic, presumably explaining our mission to the desert. At one point, the sheikh asked him a question, and al-Flayli got up abruptly. He reappeared in a few minutes, carrying a felt bag. There were gasps of astonishment around the circle when he took out a flat, circular device from the bag. He explained it to our hosts in Arabic, allowing them to pass it around the circle.

When the instrument came to me, al-Flayli switched to English. "This instrument is called an astrolabe. It dates from eleventh-century Seville."

"Amazing," I said, examining it from every angle.

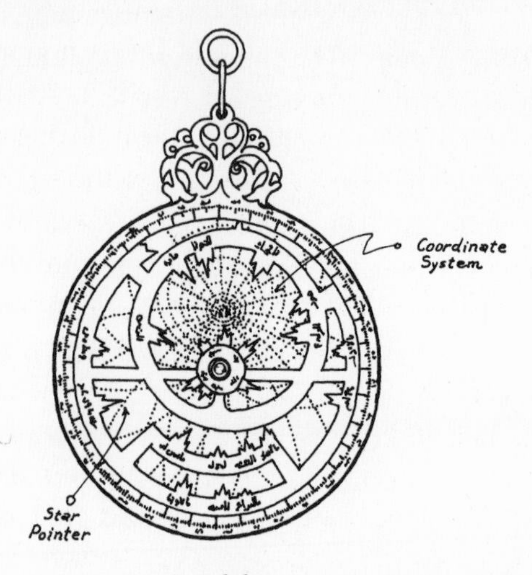

Astrolabe

It was a brass disk, just larger than my hand. On one side was a peculiar rotating circle, marked in degrees, with spokes that bore strange, jagged pointers. As I rotated the disk, the pointers moved over a curved grid of lines. Al-Flayli explained that the pointers represented important stars. On the other side of the instrument was a movable arm with a peephole at one end and a sight at the other. The arm rotated about the central post that kept the whole instrument together. The instrument was covered with elaborate writing and decorated with the most beautiful metalwork.

"As you rotate the disk on the front of the instrument, you are actually simulating the apparent rotation of the heavens about the Earth. However, the heavens reside on an imaginary sphere, as I have already explained, while the astrolabe is flat, like a map. That is why the grid lines are all more or less curved. They represent the meridians of longitude and lines of latitude on the celestial sphere."

Somehow, the instrument looked too sophisticated for the time. "Did the instrument maker calculate those lines, or did he just guess?" I asked.

"Those lines were calculated. They represent another contribution of the Arabs to the world of mathematics—*trigonometry,* or the science of angles, as you might call it. If you turn the instrument over, you will see a movable sighting arm that rotates about the center of the instrument. The arm meets a graduated scale of degrees engraved directly onto the rim of the instrument. In use, the astrolabe would be suspended by the little ring at the top, hanging straight down. The zero position of the sighting arm would then be perfectly vertical, pointing to the *zenith,* or the highest point of the heavens. By moving the arm until a given star was visible through the little peephole, you could measure the star's angular position below the zenith by reading off the angle on the rim. That angle is the one that the star makes with the vertical. It is called the *declination*.

"So there you have it all in this one little instrument. The celestial sphere has been squashed flat, mapped onto a plane, in effect. When you read an angle on the back, you merely turn the instrument over and rotate the heavens until that angle is obtained. You will then see the positions of all the other principal stars at the ends of those little pointers."

The more he explained the instrument, the more wonderful it seemed. I blurted out, "It's actually just like a miniature planetarium."

"Exactly so." Al-Flayli then turned toward the others and spoke Arabic briefly, "They are talking about the film *Lawrence of Arabia*. It was made mostly in the area we are going to. One of these men worked as an extra in the film. The sheikh himself was a little boy during the Arab Revolt."

At midafternoon we were interrupted by a youthful shout from outside the tent. I turned to see a boy astride a giant camel that plodded toward us on huge hooves, its body swaying massively. The boy held ropes that led to five other camels, dutifully following. There was something grand in the sight, something eternal.

As they approached, a man entered the sheikh's tent and muttered in the ear of the old man. "They are ready to take us," al-Flayli translated.

My plan for riding the camel was to keep my eyes shut as much as possible, only opening them by degrees. It was easy to mount the beast as she knelt. Then she rose, carrying my world of darkness aloft. I could feel her sway beneath me, not so much from side to side, but forward and back. When I peeked from one corner of my eye, I could see al-Flayli and his son, as well as two other men and the camel boy.

"You ride to the manner born." Al-Flayli drew his camel abreast of mine. "But the scenery is much better, you will find, when your eyes are open." He smiled in a friendly way, making it impossible to resent the remark.

When we had cleared the last butte, we were confronted by a sea of sand, vast dunes that alternately swallowed our little caravan in its hollows, then, on the crests, revealed other dunes ahead of us, marching to infinity.

"This is a very old trade route. This way came the caravans between Mecca and Damascus. Silks from China, spices from India, gold and ivory from Africa. When the weather was particularly hot, the caravan would halt during the day, everyone resting under awnings. They would then travel at night, under the stars. They navigated by the stars, like ships at sea. For the desert is trackless like the sea and has no signposts or roads.

"As the sage once said, 'By day thou art blind, but by night thou seest His majesty.' Tonight, Insh'llah, you will see what the desert travelers saw.

"The story of Islamic astronomy is really the story of the plane and the sphere. By the science of trigonometry, which they developed, the Arab astronomers could project the starry sphere of night onto the plane of a map, such as the one represented on the face of the astrolabe. Here is how it worked.

"Imagine a hemisphere or spherical bowl inverted over a flat surface, such as a map-to-be. Next, imagine that the bowl is covered with points in various random locations. These are the stars.

Finally, imagine a straight line hanging from each point downward to the map, touching it at a point. That point would be the position of the star on the map. Such star maps were readily produced by early astronomers, not by using threads but by trigonometric functions instead."

"Do you mean that trigonometry is equivalent to hanging a thread from each star, a thread that touches the map?" I asked.

His speech came in waves as the camel under him swayed massively forward and back. "Precisely. It was this conversion from spherical to planar position that trigonometry accomplished. Imagine for a moment that you are an early astronomer, al-Dioudni by name. Working at an advanced observatory for those times, you want to determine the position of a particular star in the night sky. To do this, you would need to measure two angles that the star makes: one vertical angle and one horizontal angle.

"For the vertical angle, your base line would be either the floor of the observatory, made dead level during construction, or, more simply, a perfectly vertical line made by a *plumb bob*, a weight hanging by a line that would, thanks to gravity, be perfectly vertical. Now you could use the sighting arm of an astrolabe to measure this angle, but because it is not a large instrument, an astrolabe is not particularly accurate.

"You would probably use an *alidade*, or sighting stick. This was actually a pair of arms made from wood or brass, fitted to a circular arc of the same material, marked off in degrees. With one of the arms perfectly vertical, matching the plumb bob, you then moved the other until you had the star in view, aligned with peepholes at both ends. You then simply read off the angle from the arc scale, and that angle would be the declination of the star in question.

"If you took the floor of the observatory as your baseline and measured from that, you would get the *altitude* of the star. Of course, the declination and altitude were freely interconvertible by subtracting either one from 90 degrees. Thus, a declination of 35 degrees is the same thing as an altitude of 90 − 35, or 55 degrees—and vice versa."

"What about the other angle, the horizontal one?" I asked. I had already imagined one possibility.

"The other angle, what today we call the hour angle, was measured by using the base of the alidade. With the star in view and the base set on another angular scale, you could read off the horizontal position of the star."

"If the baseline for the vertical angle was the line of the plumb bob," I pursued the point, "what was the baseline for the horizontal angle? Was it some fixed point of the local scenery?"

"You are right. They needed a baseline for this measurement, as well, but it was not a fixed point on the horizon. Most often, I believe they used true north, as determined by the pole star, al-Qutb. Today this star is called Polaris or the Pole Star. Polaris appears stationary, as the other stars slowly turn around it like dervishes. It just so happens that the geographic pole of the Earth itself points more or less directly at Polaris, and so it makes a natural point on which all horizontal measurements may be based. The base line for the horizontal angle would run due north, as determined by Polaris. So there you have it."

"I understand the two angles now," I said, "but how did the Arabs use those angles, and where did trigonometry come into it?"

"Well, for one thing, you could tell the time of day or night, even the time of year, using this pair of angles. If you happened to measure one of the *important* stars, those that appear on the face of the astrolabe, for example, you could look up the time in an *almanac*, an Arabic word that means table or compilation. They had inherited the Babylonian sexagesimal system of hours, essentially the one used worldwide today. Knowing these angles, you could look them up in an almanac and tell what time it was in that system of hours. You could also do the opposite. Look up the time in the almanac, and you would know where to look for any one of those stars.

"That reminds me: The word *almanac* is but one of a hundred words and names in English that have come from Arabic. Ahmed!"

"Yes, Baba?" Ahmed took a few minutes to get his camel to join ours. It was a dark little beast, with a mind of its own.

"Tell Sayed Dewdney the list of Arabic words in English."

"Albatross, alchemy, alcohol, alcove, alembic, algebra, alkali, almagest, almanac, apricot, artichoke, assassin, azimuth, azure, baksheesh, bazaar, borax, calabash, caliber, caliph, camel, camphor, cane, carafe, carat, caravan, chemistry, cipher, coffee, cork, cotton, dervish, dhow, elixir, gazelle, ghoul, harem, hashish, hazard, henna, Islam, jasmine, jinn, julep, kabob, kismet, kohl, lake (a pigment), lapis lazuli, lemon, lilac, lime, lute, macramé, magazine, marabout, mattress, minaret, mosque, muezzin, mummy, myrrh, nabob, nadir, orange, safari, saffron, sash, sesame, sheikh, sherbet, sirocco, sofa, sugar, sultan, talc, tamarind, tambourine, tariff, tarragon, zenith."

"That was excellent, Ahmed, truly excellent!"

Ahmed smiled with pleasure at his feat. "I left out the names of places and of stars," he apologized.

Al-Flayli turned at last to trigonometry: "It is such a simple idea and yet how important it is for astronomy! As everyone taking trigonometry learns, the subject concerns the angles and sides of a right-angle triangle, a subject you explored just a few days ago with your friend Pygonopolis. Trigonometry is essentially a way of translating between the angles within a right-angle triangle and the ratios of the sides. I need not stop to draw a diagram in the sand, for you can readily visualize what I am talking about by using the chalkboard in your mind. Draw there a right-angle triangle, lying on its side, with the right angle on the right. We will call the point at this angle A. Call the point above it B and the point at the other corner C."

I visualized something like this:

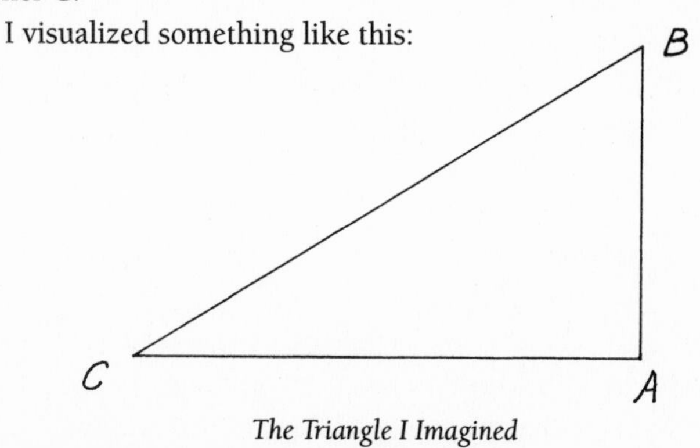

The Triangle I Imagined

"With this little system of labels, we can give names to the sides of the triangle and to its angles. For example, the hypotenuse of the triangle will then be called BC and the angle of interest, the one at C, will be called ACB.

"The so-called sine of the angle ACB is the ratio of the length of the vertical side AB to the length of the hypotenuse BC. Using the names of the sides as shorthand for their lengths, we could say that the sine of angle ACB is the ratio

AB/BC

"The other key trigonometric function is called the *cosine.* We define it to be the ratio of the base of the triangle to the hypotenuse,

AC/BC

"Now suppose that the famed astronomer al-Dioudni has just measured the angle ACB with his alidade. He would then look up the cosine of ACB, written

cos(ACB),

in a table of cosines. This is the number that represents the altitude of the star on a flat plane. On your mental chalkboard, make a circle that represents a cross section through the celestial sphere. It is a half circle on which you may place a star in an arbitrary position. Now, join the center of the circle to the star, and drop a perpendicular from the star to the floor of the half circle."

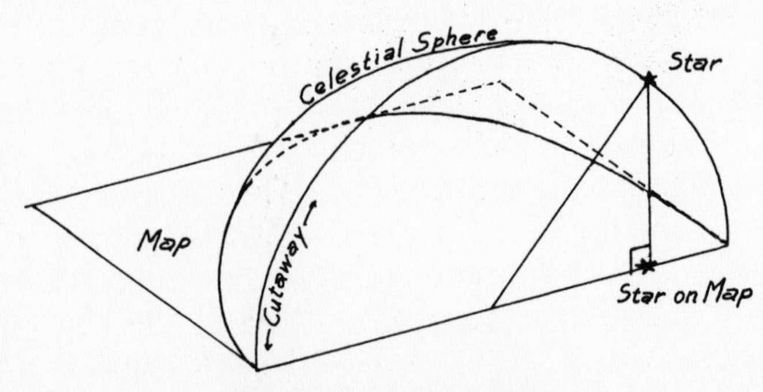

Mapping the Celestial Sphere

"Is that the 'thread' you spoke of earlier?" I asked.

"Precisely. You now have a right-angle triangle, and the ratio of the base to the hypotenuse is the cosine of the altitude angle. If we take the length of the hypotenuse as one unit, say a meter, then the cosine will be whatever the table tells us it is. Suppose, for example, that the altitude of the star is 35 degrees. Wait a minute. Ahmed!"

Again Ahmed maneuvered his camel alongside ours. "Yes, Baba?"

"What is the cosine of 35 degrees?"

"The cosine of 35 degrees," Ahmed recited, "is 0.819 to three places, Baba."

"Very good, Ahmed." Then to me, out of Ahmed's earshot: "He is a most extraordinary son. Allah has blessed us!

"So you see," he continued. "The point on the circular map we are making will be 81.9 centimeters from the center. The remaining angle, the horizontal one, can be used directly. Perhaps you can picture this map like a spider's web. The center corresponds to the zenith, and the horizontal angle coordinates are the lines that radiate outward. Along one of these lines, there is a point that is a little over 80 centimeters from the center. There is your star on the map!"

This explanation confused me until I remembered that al-Flayli had earlier spoken of a hemisphere inverted over a plane. The part under the bowl would be a disc, of course, a circular "map-to-be." When completed, the map would therefore be exactly what you would see if you could look down on the bowl from directly overhead. Just one difficulty remained.

"How on Earth did the early astronomers calculate the cosines, not to mention the other trigonometric functions?" I asked.

"I'm not sure anyone knows exactly how they calculated those tables. The simplest and easiest way, however, well within the reach of any of the early mathematicians, would be to use a kind of analog computation."

I squinted at al-Flayli to signal my puzzlement, but I could barely see him against the blaze of the lowering sun.

"I just mean they used a very large, carefully drawn circle to derive the ratio. Here's what I think they did. On a flat stone or metal surface, as large a one as possible, they inscribed a circle, precisely drawn with the finest line imaginable. Then they inscribed a diameter of the circle and marked off degrees as precisely as possible around at least one quarter of the circle. Next, for each degree, they measured the vertical length of the point above the diameter and formed the ratio of this length to the radius of the circle.

"I have done precisely this as an experiment and found that I could produce a table of sines to three decimal places of accuracy, about as good as the ancient tables. I suspect you can visualize the process."

I could, and it looked like this:

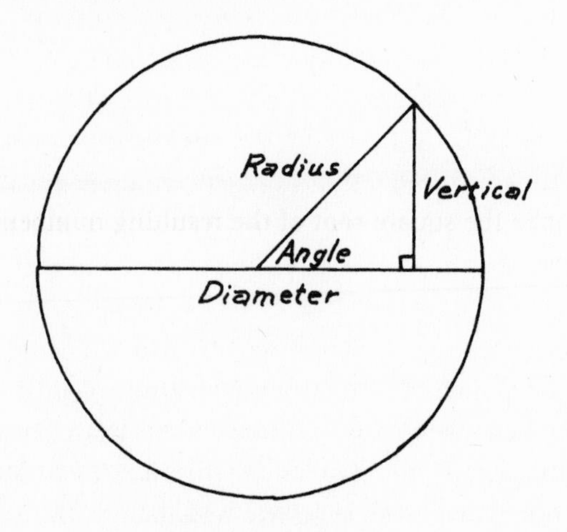

Making a Sine

"With a reasonably accurate table of sines, the other trigonometric functions could be derived by arithmetic. I can illustrate this by returning to the right-angle triangle we discussed earlier, the one we called ABC. The sine of the angle ACB is the ratio of the vertical side, AB, to the hypotenuse, BC, while the cosine is the ratio of the floor, AC, to the hypotenuse. Now, let me recite the Pythagorean formula:

$$AB^2 + AC^2 = BC^2$$

"All we have to do is divide all three quantities by the square of the hypotenuse to get

$$(AB/BC)^2 + (AC/BC)^2 = (BC/BC)^2$$

In other words, because the ratio BC/BC is just unity or 1, the squared ratios turn out to be a squared sine and cosine.

$$\sin^2(ACB) + \cos^2(ACB) = 1$$

Finally, to get the cosine from the sine, we use a little more algebra to get first

$$\cos^2(ACB) = 1 - \sin^2(ACB)$$

and then

$$\cos(ACB) = \sqrt{1 - \sin^2(ACB)}$$

"In other words, if you happen to know the sine of an angle, you can easily get the cosine: You square the sine, subtract it from 1, then take the square root of the resulting number. That will be the cosine.

"For mathematicians, it is an almost trivial formula, yet it illustrates an important point about mathematics: Knowledge from many different areas of mathematics flows seamlessly together. There are first the definitions: those of the sine, cosine, and other trigonometric ratios that arise from a right-angle triangle. Then there is the Pythagorean theorem, which also concerns right-angle triangles. Then there are the rules of algebra by which the formula was derived. Finally, there are the applications, actual numbers, the sines of angles that you might want to plug into the formula to find the corresponding cosines. Most formulas arise in such a manner.

"Perhaps my imagination is lacking," he concluded, "but I can't imagine a shorter or simpler formula for converting sines to cosines. As for the sines, they must have been measured!"

"What some people may not appreciate," I ventured, "is that the accuracy of those measurements would depend critically on the accuracy with which the circle was drawn."

"That is true. Come to think of it, that observation leads to yet another answer to your question concerning the independence of mathematics and the problem of discovery."

This had not occurred to me.

"Does it not tell you something, this phenomenon? To the exact extent that the precisely drawn circle is the embodiment of the abstract circle, the circle from the Superior World, the results will be precise. Two quite different people compiling a table of sines by the method I outlined might well get slightly different results, at least in the last digit or so. Yet as each tried a succession of ever-larger and more finely drawn circles, not only would their tables agree, but also the precision of both their results would improve apace. They would be discovering the actual values of the sines."

"But that's a little different, isn't it?" I responded. "After all, they are working not on abstract circles but on material embodiments of them."

"Precisely. For that very reason, you can at least see and touch the source of their information. The errors they make will arise from slight imperfections of the drawn circle, the rulers they use to measure lengths, even the way they make the measurements. In fact, errors of instrument and of observation creep into astronomy no less today than a thousand years ago. Today, however, we have a mathematical theory of errors, a branch of statistics, that tells the extent of the errors so that we know how close we are to a correct answer. Can we even conceive of such a possibility without there being something behind it all, without there being such a thing as a correct answer? And where does that correct answer exist?

"If you like, you may imagine that the correct answer exists in the circle as drawn. But who is so dull of wit as to be incapable of imagining the ideal circle to which the material objects appeal, in effect? Who cannot imagine the —"

"Baba, we are here!" It was Ahmed. We had reached a larger-than-normal hollow between two low dunes. The men of the tribe knelt their camels, which roared and groaned as though they had walked a thousand miles. Bags were thrown open and tents pitched. The shouts of the men were lost in the immensity that surrounded us. They lit a fire and began to cook the evening meal. The sun was already behind the dune to the west.

I cannot remember a more delicious meal. The men had cooked lamb over the fire and made a sort of gravy. After the sunset prayers, the cook had strewn pieces of lamb over a large, flat brass pan lined with something like pita bread. Then he had poured gravy over the lot. We knelt around the pan in a circle, eating in the traditional style.

"I wanted you to get at least a taste of life in a caravan," said al-Flayli over the meal. "The Arabs—no, what am I saying?—the Semites ate like this for many thousand years, back to the time of Abraham and beyond."

After supper, the night grew completely dark. We sat around the campfire on cushions, the brightness of the fire extinguishing all sight outside the magic hemisphere of its light. The camel boy circled around the perimeter, carrying a brass pitcher from which he poured an ablution for each of us. We washed the grease from our hands and watched the rinse sink into the desert sand. Al-Flayli gave an order, and the men doused the fire.

"Now look up!"

We were inundated by an immense and sparkling majesty, more stars than I could ever count, scattered like jewels across the heavens. Such metaphors came readily enough at the sight. It was, in a word, mind-boggling.

"Now you are seeing what the early traders saw as a nightly matter. Now you know why they were so familiar with the sky and why so many stars were named by them.

"Ahmed. If you please, recite the names of the stars." Al-Flayli leaned toward me. "I mean all the Arabic names, as used by astronomers today. It might be asked, what is an 'astronomer' but one who names the stars?"

Once again, Ahmed happily recited: "Achernar, Aldebaran, Algol, Alioth, Alkaid, Almach, Alnath, Alpharatz, Alphard, Alphecca, Alsuhail, Altair, Antares, Arcturus, Betelgeuse, Caphe, Deneb, Denebola, Dubhe, Etamin, Fomalhaut, Hamal, Kochab, Marfak, Mirak, Mizar, Phecda, Raselague, Rigel, Schedir, Shaula."

Al-Flayli smiled at his son, then turned toward me, glancing upward. "Look up and feel the spell of the grand illusion. The stars are beautiful, to be sure, and some are brighter than others, but do they not all reside on the surface of a huge sphere? We can see only half of it, the other half lying below the horizon. But clearly it is a sphere. Everybody today sees it that way, even professional astronomers who know that some stars lie much, much farther away than others. There are even some bright stars that lie a hundred times farther away than other, very faint ones nearby. You see, purely intellectual knowledge is sometimes a very different thing from experiential knowledge. The most difficult thing in the world is to know when to set one kind aside in favor of the other.

"The point is, you cannot by an act of will set aside this most natural of perceptions, that the stars all reside on the surface of a vast sphere. There it is above us for all to see!

"The ancients were no less persuaded of this simple and obvious fact—so much so that the Greeks maintained that the stars were fixed to the surface of a vast sphere beyond which lay Olympus, home of the gods. It was just as natural for the Arabs, too, to see the sphere. Today, we call it the celestial sphere, a handy fiction that is useful only for locating the position of stars in the sky, without reference to their distances from us. Ah, the distances from us!

"As it says in the Koran, 'By the stars and their places, if you but knew what that means!' If that was a hint from the Koran that the celestial sphere was an illusion, the Islamic astronomers missed it completely.

"The sphere was a symbol of perfection in three dimensions, just as the circle was a symbol of perfection in two. Was it not natural for Allah to arrange the stars on the surface of a sphere? Now, as the ancient astronomers looked more closely at the night sky, they realized that there must be more than one sphere up there. This, in my opinion, was the beginning of true cosmology. The planets, for example, obviously did not belong to the sphere of the stars because they wandered freely in front of it. The ancients not only knew that the planets must be closer, but also thought the planets must be on a different sphere—or spheres.

"In keeping with the magnificence of creation, therefore, early astronomers conjectured that each planet lay on its own sphere. They knew of five planets, from Mercury to Saturn, each carried about the heavens on its own revolving sphere. The sun was carried on yet another sphere, making a total of six. Very quickly, such conjectures became accepted fact. It remained only for astronomers to account for the wanderings of the planets by using spheres, and only spheres, to explain their motions.

"The interesting thing about all this is that it represents a kind of groping toward the truth. It was true, after all, that the planets did not belong to the celestial sphere. They are much, much closer to the Earth than any star."

I was shivering, half from the excitement of sitting in this natural amphitheater of the stars and half from the increasing cold. The camel boy brought blankets out for everyone at just the right time.

"Baba," Ahmed interrupted, "what about the House of Wisdom?"

"Later, Ahmed, later. I am talking about the spheres of the heavens and I am about to make a most important point for our guest. As far as the Islamic astronomers were concerned, Allah had arranged that the celestial sphere rotated about the Earth once every day. The problem was to assign a system of coordinates to this sphere so that the positions of stars on it could be mapped with precision.

"At first, they probably tried a horizontal system. For this, they needed two fixed points—the zenith and true north, as I mentioned before. This choice supported the two angles—declination and the angle from true north. But every night, even at the same time, these stars shifted ever so slightly from their previous position. As the seasons succeeded one another, the pole star would appear to move first toward the south, then back toward the north again, completing the cycle in one solar year.

"The pole star, being the axis of the celestial sphere, caused the whole sphere to follow the same annual progression. Would it not be more natural to adopt a system based on this simple observation? A newer, equatorial system of coordinates was devised. In this system, there were also two coordinates. One coordinate was essentially the same as before, the declination not from the zenith but from Polaris. The other angular measurement would have to be at right angles to this, along the celestial equator, a great circle in the sky that corresponds to the Earth's equator."

"The baseline for this second coordinate had to be carried around by the celestial sphere itself. It had to correspond to one of the stars on the celestial sphere. Once that point was chosen, they had a system of coordinates of the stars, which did not change from one day to the next or from one month to the next. They remained always the same.

"The Arab astronomers could now compile a new, much simpler and smaller almanac with but one set of entries and no reference to time at all. They also knew how to convert from one system to the other. To achieve this they needed the older-style almanac for only two stars: Polaris and the base star for the new equatorial coordinate. Suppose they wanted to know what position Aldebaran would have on a given night at a given time. They would look up the positions of these two stars for that time on the old almanac, then they would look up the position of Aldebaran in the new equatorial almanac. Then all they had to do was add Aldebaran's coordinates to those of the corresponding base stars in the old almanac, and they would be done. If they erected a pole that pointed to the celestial sphere with those precise coordinates at the time in question, the pole would point directly at Aldebaran."

Al-Flayli produced a flashlight from his pocket. "But I do not want to talk about the stars now. Indeed, let us turn to the planets."

When he switched on the flashlight, we could see a thin but definite pencil of light mount heavenward. With this beam, he could sketch ideas on the sky itself, a starry chalkboard. Interestingly, no matter what star he pointed at, everyone saw the beam point to the same place. Everyone knew what star he meant.

"There is the pole star behind us." The flashlight made smooth circles around the pole star, circles that widened until they intersected the horizon. "All these circles represent the tracks that stars made in the sky every night."

"The celestial equator looks like this." The flashlight started somewhere to our east, swept up into the sky in a grand arc, only to sweep down to Earth to the west. "Unfortunately, I can only show you the visible half of the celestial equator. The other half lies on the other side of the Earth. And now look, here's another equator of sorts."

He swung the beam of the flashlight along a slightly different great circle this time, naming constellations as he went: "Scorpio, Sagittarius, Capricorn, Aquarius, Pisces, Aries." He stopped twice along the way. "Look, there's Mars, that reddish point of light over there. And this creamy-looking light is Saturn.

"These are the constellations of the zodiac, so named because the sun and planets all pass through them in the course of a year or more. In reality, this great circle is called the ecliptic. It represents our edge-on view of the plane of the solar system. And because the sun and planets all lie on this plane, they will always be found somewhere along the ecliptic, just as Mars and Saturn may be found there tonight. The ecliptic provided yet another great circle, the foundation of yet another coordinate system.

"The planets produced a wonderful income for some of the astronomers, who were also astrologers. But they also produced endless trouble for almost any scheme proposed to account for their motions. If each planet followed its own sphere, there was something very peculiar about those spheres. Long before the Arabs, Greek astronomers like Ptolemy and Apollonius had noticed that the planets, even as they moved smoothly and incrementally night after night along the ecliptic, would occasionally double back on their own paths, a most puzzling phenomenon."

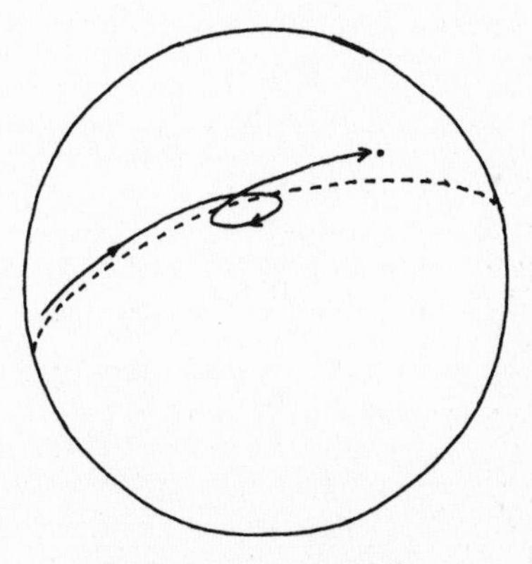

A Planet's Apparent Path

Al-Flayli waved his flashlight slowly across the sky, moving it in tiny jerks.

"What astronomers would

observe, as they measured the position of a planet on successive days or weeks against the celestial sphere, was a path that would sometimes speed up and sometimes slow down. Periodically, the planet would even reverse its course, performing a backward loop.

"As far as Apollonius and Ptolemy were concerned, the only way to preserve the perfection of the heavens was to explain these odd movements as the result of a second sphere, rolling within the first. It was called an *epicycle* and worked like this: What I will trace now is the path of such a planet, as seen from above the orbital plane. I will merely use the sky as a kind of celestial chalkboard."

Al-Flayli traced a roughly circular path in the sky with loops in it. It traced the effect of one sphere rolling within another, reminding me of epicyclic diagrams in ancient texts.

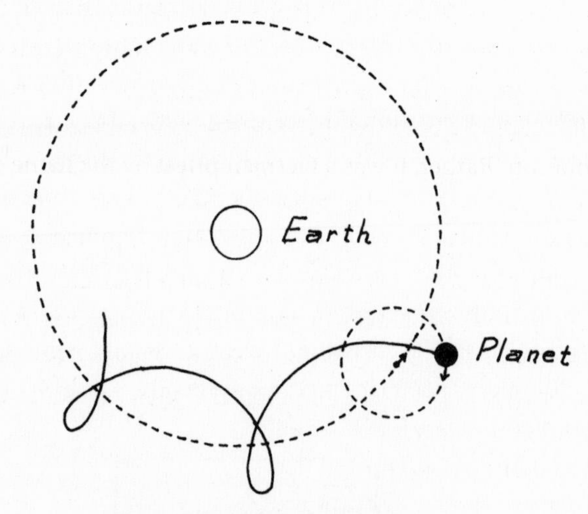

A Planet Moving in Epicycles

"When Islamic mathematicians plotted the effects of such motion on their celestial spheres, the resulting tracks were in close agreement with what they observed, at least in enough cases to convince them that Ptolemy's theory was correct. Indeed, Ptolemy was part of their received wisdom from the Greeks. So much else in Greek science was so clearly true, not to say wonderful, that they had little reason to question the Ptolemaic theory.

"This theory held sway for over a thousand years. Only toward the middle of the Islamic era, around the twelfth century, did Muslim astronomers begin to question the Ptolemaic theory. They saw too many discrepancies with their observations. Al-Tusi, the chief astronomer at the Maragha observatory, ceased to believe in the Ptolemaic system, proposing a new one of his own, but it was essentially a variant of the epicyclic theory. The Spanish astronomer ibn Aflah publicly criticized the Ptolemaic theory, as did others. They knew something was wrong but couldn't quite put their fingers on it. Only during the European Renaissance did astronomers discover the true state of affairs.

"The so-called Copernican revolution consisted in a proposal that put the sun at the center of the cosmos, with the Earth and other planets going around it in circular orbits. The heavenly spheres made little music, but perhaps only a faint tinkling of breaking glass, as they disintegrated in the minds of men.

"In truth, Copernicus is not the real father of the so-called revolution. Rather, it was a German priest by the name of Johannes Kepler. At first, Kepler more or less ignored the Copernican theory. He worked for many years on the problem of determining the sizes of the spheres that carried the planets around the Earth. He finally hit on a scheme of great elegance and beauty. Each sphere was determined by one of the so-called Platonic solids, progressing from the cube to the tetrahedron, then to the dodecahedron, and so on.

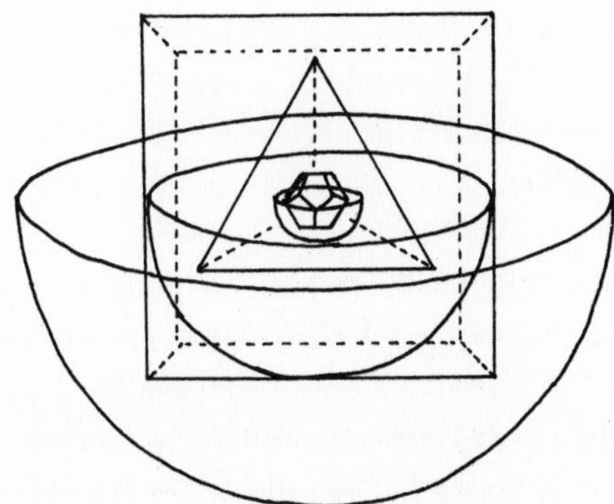

The Mysterium Cosmographicum

"He felt an almost mystical rush of emotion at this discovery, feeling that he had penetrated to the very heart of the cosmos. Kepler, you see, was a Pythagorean and held strongly to the view that the ultimate answer to the riddle of planetary motions would lie in mathematics. It was an article of faith with him no less powerful than his Christian convictions.

"But the actual motions of the planets did not fit this wonderful scheme, and Kepler reluctantly abandoned it. Only then did he turn once again to Pythagoras for inspiration. The circle, proposed by Copernicus and one of the conic sections, had not been a success. After many years of painful calculation during which he nearly starved, he finally hit on the idea of trying another conic section, the ellipse.

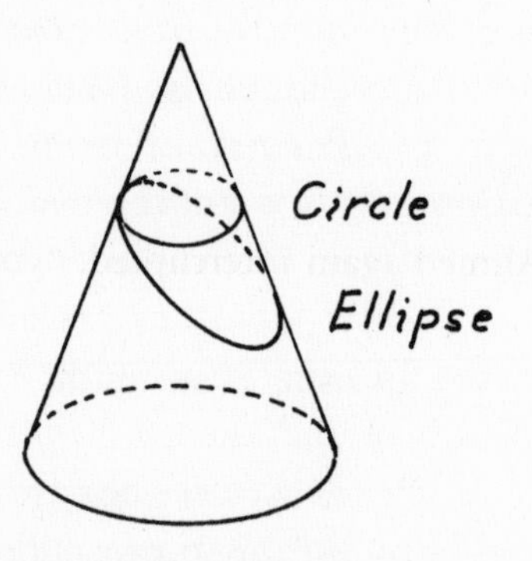

Two Conic Sections

"There was much calculation to do as he compared actual planetary data with the predictions of a theory that had the planets travel along ellipses around the sun, instead of the Earth. The match was breathtakingly good. Pythagoras himself would probably have been delighted.

"Let us not forget that the whole meaning of the Copernican revolution lay in the revelation of falsehood. The earlier astronomers had been wrong. It is true that one worldview replaced another, but the second view was driven less by the wishes of men than the first. The Renaissance freed them from the need to preserve the spherical structure of the heavens. As I have explained, some Islamic astronomers already knew perfectly well that the Ptolemaic model was simply wrong. But they could not, by a mere act of will or imagination, devise the correct model, for it meant giving up cherished notions of heavenly movement. If you want to say that culture influences science, I shall not argue with you. But paradoxical as it may sound, only the possibility of being wrong will save science from becoming a purely cultural exercise.

"The Copernican revolution was actually a rescue operation, whereby a superior model, one that we now know to be correct, replaced an incorrect model, one inspired by the spherical illusion of the heavens and preserved thereafter by purely cultural influences. Don't forget, the cosmos is no longer distant from us. We are out there daily in our spacecraft, computing their orbits with a precision that is good enough to rendezvous within a meter after a journey of hundreds of thousands of miles. This simply would not be possible if the modern view of cosmic dynamics were in any serious way incorrect."

"Baba," Ahmed again interrupted, "you said that you would finish —"

"Yes, Ahmed. It is time. If our guest does not object, we will return to the House of Wisdom.

"The year is A.D. 926, the year 211 of the Islamic era. Al-Khwarizmi stands in the courtyard of the Bayt al Hiqma with Caliph al-Ma'mun. The stars are out, and the caliph is lost in wonder. 'Tell me, O wise one,' he addresses al-Khwarizmi. 'When the philosophers say "As Above, So Below," what mean they?' Al-Khwarizmi replies, 'Many things, O Shadow of Allah. I will illustrate but one. There to the north you will see the pole star, al-Qutb, about which the heavens turn every night. As your eminence can see, the heavens are arranged as a vast sphere, according to divine plan. As above, so below.

"'Of this principle, O Caliph, we see an apt illustration at our very feet: If the heavens are a sphere, so is the Earth. We have measured the angle of al-Qutb here in Baghdad to be 33 degrees and 19 minutes above the northern horizon at the beginning of spring. Whereas in Mecca, where is found the House of Allah, we measure 21 degrees and 33 minutes at the same time of year. Now why should there be such discrepancy? As we go from Baghdad to Mecca, the angle of al-Qutb changes by nearly 12 degrees of arc. How else to explain this but that we live upon a sphere, the very reflection of the heavens themselves. As we cross the land and observe the pole, it drops by successive amounts the degrees of latitude, however small, that we crossed each day. Verily, we live upon a sphere, we —' "

"Ahmed! What is wrong?" Al-Flayli suddenly cried out.

Ahmed was crawling on hands and knees toward the tents.

"I can't look up, Baba. I must go inside," Ahmed responded in distress.

Al-Flayli relit the fire. Against the glow of new flames, he shrugged.

"To some people is given the ability to shrug off the illusion of the bowl. Ahmed is such a one. Sometimes, when he looks up, he sees not the bowl, but into the depths of space. The perception is terrifying beyond measure. It is perhaps time for all of us to retire."

With that, I left for the tents while al-Flayli and the others said the night prayer. Ahmed did not come out to join them, choosing to pray in his tent instead.

PART III

THE VANISHING ACT

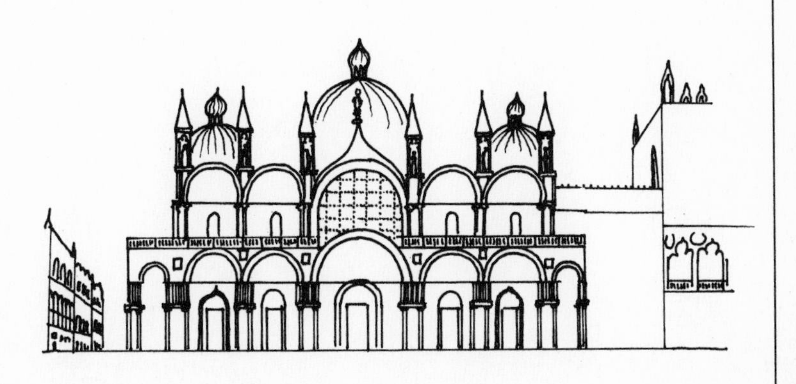

CHAPTER 5

The Message

Venice, Italy, June 26, 1995

I said goodbye to al-Flayli on the tarmac of the airport in Aqaba. Just before I stepped aboard a perilous-looking old twin-engine aircraft for the flight to Cairo, al-Flayli said something rather odd. "I'm not sure you realize how important your quest is. In the philosophy of science, indeed in the practice of science, there is no more fundamental consideration."

I pondered these words as the plane flew over the Sinai desert, the Suez Canal, and on to the bustling Cairo airport, where I would change planes for the flight to Venice. Sitting amid a row of suitcases on which sat several mothers and infants, I had time to catch my breath, philosophically speaking. What had I learned from my visit to Aqaba?

While it was true that the evidence, such as it was, piled up in favor of the independent existence of mathematics, as well as the reality of its discovery, I had been presented for the second time with what might be called a grand failure. Pythagoras had thought that all distances were commensurable, then he had discovered that he was wrong. The second example of failure, the celestial sphere, was not really a mathematical failure but a scientific one. Ancient astronomers from many different lands and cultures had all seen the stars affixed to a rotating sphere. At base, the opinion (it could hardly be called a hypothesis) was surely the result of a powerful visual illusion. Each culture—including medieval and very Christian Europe—had seen its fundamental beliefs fulfilled in the celestial sphere. Indeed, it appears to have been the role of culture in each case to reinforce the illusion.

There were ironies playing amid these cross-currents of mathematics and culture. It seemed that so far, culture, far from producing differences, was producing similarities, whether a theory was right or wrong. Also, how ironic that Kepler, the self-professed Pythagorean, should be the one to finally discover the actual motion of planets! Before this discovery, however, even Kepler had taken a last, passionate stab at the spheres, rejoicing finally in a vision of the *mysterium cosmographicum,* in which the spheres became transparent Platonic solids nested about a stationary and central Earth. In any event, observation had not agreed with spherical (i.e., circular) motion, whether around spheres or Platonic solids. Nor had it agreed with the predictions of epicyclic motion. Observation agreed with the new Keplerian laws perfectly, however. That could be no accident.

These thoughts acquired such momentum that I barely noticed the flight over the Mediterranean, landing at the Venice airport with a new sense of anticipation. What strange and wonderful things would I learn from my next contact, Maria Canzoni of the Universita' Ca' Foscari di Venezia.

Canzoni was not at the airport to meet me. Instead, I found my name, rather strangely rendered, adorning a placard that waved above the crowd around the arrival gate. it was held by someone who introduced himself as Emilio, a student of Canzoni's. He chatted cheerfully as we drove along the slender causeway to the ancient city of Venice, apparently afloat on the Adriatic Sea. The water looked murky.

"Is that water clean?" I asked.

Emilio looked at me as if I were from Mars. "You fall in there, you die." He laughed. We parked the car, then took a water taxi to the university.

Canzoni was young, with prematurely gray hair and a gracious, almost reserved manner. Despite her reserve, she took my hand and held it while she explained her rudeness in not showing up at the airport.

"We had a committee meeting. It was about two new graduate students—whom to accept and whom to reject. A painful business. It was supposed to end two hours ago. And how was your flight? Have you been to Venezia before?"

"Great, thank you, and no. I had no idea what an amazing and beautiful city this is. You're lucky to live here."

"Yes", she said matter-of-factly. "Very lucky."

We entered a converted palazzo where Canzoni had an office. The front door said "Dipartimento della Historia." She seemed to relax once she was seated behind her desk. I took a plush chair, sat down, and promptly yawned. Now it was my turn to apologize, explaining that I had spent the previous night camped in the desert.

"But how exciting. How lucky you are! The desert is much more exciting than this place, I think! And was your trip to the desert in search of the reason for why mathematics is so powerful?"

I began to tell her about al-Flayli and al-Khwarizmi, at the mention of which Canzoni smiled and leaned forward, as if confiding a secret.

"I am sure you know that we get the word *algorithm* from that name. It used to mean simply 'methods of calculation with numbers.' It was introduced to Italy and Europe by Leonardo di Pisa, also known as Fibonacci. He was not really a mathematician, at least not at first, but a trader who worked for his father. He traveled frequently to Cairo, Tunis, Algiers, and other North African ports. He was impressed at the ability of his Arab counterparts to calculate bills of lading, prices, values, even weights and measures. Their methods looked especially efficient when compared to the clumsy Roman system still used in medieval Europe.

"He persuaded his commercial contacts to teach him this numerical craft, and he proved an apt pupil. To be completely sincere, it may be that he was motivated not only by the noble idea of spreading useful ideas but also at seeing his name become a household word. He would write a book, and it would become indispensable reading to every merchant, every banker, every tax collector, even every monk! He knew that his aim would be achieved if he wrote clearly and well, using many examples of every kind. The book he wrote was called, simply, *Algorismus,* offering a Latin title for tone, suggesting a mystery, we could say."

She smiled sadly, as if remembering a long-vanished happiness. "Did I mention in any of my e-mails to you that I began my career as a physicist? My background has made your questions about the reality of mathematics all the more interesting. Who but a physicist works daily with a foot in both worlds, the physical and the mathematical! Knowing that you would visit has made me think extra hard, and I have some good examples for you, perhaps."

She told me about her time at the large particle-research facility at CERN in Geneva, the center of European physics. While searching her books and papers for examples of the unreasonable power of mathematics in physics, she had inadvertently begun to relive that period of her life. I had the impression that something had happened to her at CERN, something that drove her away. Or had she come to Venice not as a fugitive but as a pilgrim? Did she then merely regret something?

"I propose to produce for you a kind of vanishing act in which, as we examine physical reality ever more closely, it gradually fades from view, replaced by something quite different." She sighed, "Yes, quite different!"

I hinted as strongly as I could that she could begin to tell me about it.

"You must permit me something of a buildup. You may ask me directly whether mathematics exists independently, or you may disguise the question, pretending it is about discovery versus invention. But they are the same question! How can they not be? If you ask such questions, I will answer by describing the great vanishing act. But please, you must permit me some other examples in case my grand thesis does not satisfy you."

As she talked, she gathered some things and put them in her briefcase. We were going to dine out, apparently. I was famished.

"After the dinner, at 7:30 this evening, there is a special recital of Gabrieli at San Marco. I asked myself, 'What better introduction could there be for you to the very spirit of Venice than to hear Gabrieli's "Missa Sancta"?' It will be performed in the very church where he worked as organist and master of the choir."

The stones of Venice, it is said, have a sweet smell of their own. Most tourists complain of the stink from the canals, and a certain dank and fruity sweetness seemed to follow us along the walk to the Rialto, with its covered bridge. If I closed my eyes to mere squints, the people all changed into medieval dress: traders, brokers, hawkers, vendors, investors, and their ladies. It was Fibonacci's world, recently quickened by the introduction of *Algorismus*.

We climbed a set of stairs and found ourselves amid the delicious smells of a wonderful little *ristorante* tucked away on the second floor of an ancient building, with a *terrazzo* that overlooked one of the lesser canals. We could not see the Adriatic from the *ristorante* but could feel its warm, humid breeze wafting along the medieval canyon that enclosed us. We dined on scampi and mussels in a rich, cream-laden pasta. As soon as the waiter took away the plates, Maria reached into her briefcase and took out several papers.

"There is just time before we leave for San Marco to discuss some matters pertaining to your questions. One of the best examples I found was the work of a Swiss mathematician named Balmer." She handed me a paper.

"And what was his work an example of?" I asked.

"It demonstrates how mathematics inheres in matter. You could say that this one example shows that mathematics, as an entity, is at least as real as the so-called real world, perhaps even more real."

I looked down at the paper while Canzoni signaled the waiter.

BALMER'S PROBABLE MODE OF SOLUTION OF THE HYDROGEN SERIES PUZZLE

Maria Canzoni,
Dept. of History of Science
Universita' Ca' Foscari di Venezia
Venice, Italy

ABSTRACT

In 1884, Johann Jacob Balmer was a teacher in a girls' school in Basel, Switzerland. Trained as a physicist, and a very good one, Balmer had never succeeded in ascending the academic ladder beyond the position of *privatdozent* [tutor] at the University of Basel. In 1884, however, Balmer was to do something quite astonishing. He had puzzled for months over some observations reported by Angstrom. These observation took the form of four 6-digit numbers, the wavelengths at which Angstrom had measured strange lines in the spectrum of hydrogen gas. Physicists of the time found it very remarkable that hydrogen should absorb and emit energy only at specific wavelengths. Balmer discovered a two-variable formula with values that matched these wavelengths to within the experimental error of 1 part in 7,000. The existence of a formula for the discrete spectral lines of the hydrogen atom led directly to the discovery of quantum states of matter.

I understood much of this. Anders Ångström was a Swedish physicist and one of the first people to use a new instrument called the *spectrograph,* which broke light into an array or spectrum of component colors, like a prism. Each color in a spectrum represented light of a specific wavelength. Scientists had expected that all spectra would turn out more or less like ordinary sunlight, composed of a continuous band of wavelengths. Ångström and other physicists had been much puzzled when they examined the light emitted by certain hot gases, such as hydrogen. Instead of a smeared-out, continuous spectrum, they found discrete lines at specific wavelengths.

Spectra of Sun (above), Hydrogen (below)

As most people know, light from the sun, which appears yellowish-white, is composed of a nearly continuous rainbow of wavelengths, from infrared to ultraviolet and beyond. In contrast, hydrogen gas does not show this continuous spectral array. When heated, it has its own characteristic color, a strange shade of violet. When this color is viewed through a spectrograph, it turns out to consist of discrete lines, one for each wavelength composing the color. It was later discovered that each wavelength arose from a different state of vibration of the hydrogen atoms in its gas.

Balmer's formula reproduced the wavelengths of excited hydrogen. It used two integer variables, n and m, which independently took on small values such as 1, 2, 3, and so on, producing all the numbers found by Ångström, and then some. I glanced ahead in Canzoni's paper and found the formula. No doubt, she would give it a fuller explanation.

She leaned over somewhat anxiously. "You will find an important little table on page 37." I turned to the page and found Ångström's original measurements.

$$H_\alpha = 6562.10$$

$$H_\beta = 4860.74$$

$$H_\gamma = 4340.10$$

$$H_\delta = 4101.20$$

Canzoni explained their significance. "The wavelengths were given in the units that would later bear Ångström's name: one ten-billionth of a meter. The numbers were the wavelengths of the first four lines in the hydrogen spectrum, as measured by Ångström, who had a reputation for extremely accurate work. Balmer was convinced that the numbers had a very special interpretation, one that would justify the school of philosophy to which he belonged."

I gave Canzoni a quizzical look. If I'd had any idea of what she would say next, I would not have taken such a large sip of my Brio Supremo.

"The school of Pythagoras," she said.

Canzoni looked greatly concerned when I began to cough and wheeze, fighting desperately to expel the drink away from my lungs. Diners at other tables looked askance at us. When I had recovered sufficiently, I explained how that name had come up in my previous two visits. She immediately wanted to know more about both. She took a particular interest in Pygonopolis, clapping when I came to the holos.

"I must get his address from you," she cried. "It is important that we have a name for this place, and in matters classical, we Italians must all too often bow to the Greeks! The holos, the holos, the holos. I like it better than the Platonic World. The holos and the cosmos. Beautiful!"

By the time I had explained the ideas of both Pygonopolis and al-Flayli, hardly more than sketching them, it was nearly eight o'clock.

"Come. Come. Come." She had been drinking just a bit more wine than was perhaps good for her. We caught a water taxi at the canal by the restaurant. The wind caught her hair as we stood watching the lively scenes of gondolas and motorboats. She turned to smile solemnly at me.

"You are right. There is no city on Earth with half the beauty of this one."

At the piazza of San Marco, we strolled to the magnificent medieval church, only to discover standing-room only inside. The concert had already started, so we stood at the back, listening to the magnificent kyrie reverberating from saint to saint on the lofty golden ceiling. Canzoni motioned me outside, saying that she doubted she could stand that long. She had the *scleroso,* which I suddenly realized meant "multiple sclerosis." Once on the bench, she listened intently to the music filtering from the ancient church.

"It is beautiful, but I must tell you something," she whispered. "Such music spelled the doom of the Church."

"I would have thought just the opposite!" I remarked.

"We do not understand religion today the way they understood it in the medieval period. You see, by expressing one person's religious awe, in this case that of Gabrieli, such music made worshipers dependent on it, robbing them of their own inner music, so to speak. Instead of lighting spiritual candles, the music extinguished them. This idea, which for some reason I feel compelled to share with you, is held by almost no one today, yet it is there in the sources, in the trepidation expressed by the church fathers. Compare the richness pouring from that door with the simplicity and purity of a Gregorian chant."

Her comment about religion reminded me of Pythagoras and the Pythagoreans. The revelation of Balmer's Pythagorean learnings had truly startled me. My quest had developed a leitmotif that would not go away. To find Pythagoras in the year 500 B.C. was one thing, but to find him again among the Middle Eastern Brethren of Purity in A.D. 900, then popping up in Kepler's sixteenth century and again in Balmer's nineteenth was surprising, to say the least. I felt compelled to ask, "What do you know about the Pythagoreans?"

"The Pythagoreans were mystics, technically speaking, with a chain of initiation that went back to Pythagoras. Indeed, it probably went back even further to Thales and to the master of Thales, a mysterious character by the name of Berossus of Babylon.

"To be frank, we know very little else about the practices or doctrines of the Pythagoreans, but they believed that contemplation of the Platonic world, the holos, brought them closer to the 'Uncreate Source,' what today we call deity. We have only a few tantalizing hints of how the sect operated. They wore white, for example. They took vows of purity and rectitude. They had a blue star tattooed on the palm of their right hands; they were sworn to secrecy on all matters of doctrine and were even forbidden to eat beans; and so on. Our knowledge consists of little fragments such as these."

I wondered out loud whether there had been something like a Pythagorean sect operating in nineteenth-century Basel.

"Frankly, I doubt it," said Maria. "Balmer was probably a romantic who found in descriptions of Pythagoras and the Pythagoreans a repository of his own feelings about physics and mathematics, and the relationship between them.

"This would mean that Balmer was very much on the lookout for physical theories that could be based on integers or ratios of integers. As a Pythagorean, even a romantic one, he very much believed that all physical theories would one day be based on the integers. He believed, in short, that the cosmos was discrete, broken up into fundamental units that reflected a corresponding structure in the holos. How handy that word is!

"With the rise of the Daltonian atom as a hard little sphere and its subsequent elaboration by Rutherford and Dirac into further discrete structures such as nucleus and electrons, Balmer saw his own view of things justified. A discrete universe was beginning to make its presence felt. In any event, I am getting ahead of myself. I will say much more about these developments tomorrow.

"By the 1890s, the spectrograph had revealed a new world of wavelengths. Most interesting of all was the light emitted by hot gases of excited atoms of various kinds. These invariably showed not continuous rainbows, so to speak, but just a few lines, a few wavelengths that characterized the atoms in question. Such spectra were called *emission spectra* because they broke apart the light emitted by atoms. There were *absorption spectra,* as well: If you looked at a light source, such as the sun, through a gas of specific atomic composition, you would see dark lines at the same wavelength positions are you would in the emission spectrum for the same gas.

"This meant that you could detect all sorts of elements in distant stars, or even in near ones, such as the sun. Although the hydrogen lines would be more or less lost amid the wealth of other lines in a solar spectrum, essentially the rainbow, you could still detect gases such as hydrogen by passing the sunlight through hydrogen gas. The appearance of dark lines for hydrogen in solar and stellar spectra, for example, could only mean that hydrogen was present in these bodies. If these wavelengths were not emitted by the source, how could they be absorbed?

"Thus, when astronomers examined bright stars such as Altair or Deneb through the new instrument, they were surprised to find large amounts of hydrogen gas. A number of physicists began to measure these lines very carefully, although none with more care than Ångström. In fact, his measurements of the wavelengths were amazingly precise, to better than 1 part in 7,000. Ångström published his data, and many scientists puzzled over the mysterious numbers that emanated from the stars.

"When Balmer's eyes first fell on those numbers that Ångström had so carefully extracted from the cosmos, he turned to the holos for inspiration. The wavelengths found by Ångström looked superficially as though they might be irrational, containing an infinite number of digits, yet Balmer's Pythagorean faith held firm. Surely there were integers lurking within those mysterious six-digit messages from the stars!"

Canzoni was obviously heading toward Balmer's solution of the enigma. I said nothing, letting her talk, but I wondered how anyone might tease integers out of such murky-looking numbers without resorting to mathematical chicanery.

"Balmer undoubtedly tried many approaches. For example, he probably tried to take ratios of the wavelength numbers directly. First, let us list the four numbers that Balmer worked with, as received from Ångström. The hydrogen spectrum, you see, consisted of a principal or alpha line (H_α) in the red region of the spectrum, then a secondary, beta line (H_β) in the violet, then a third, gamma (H_γ), and a fourth, delta (H_δ), and so on. At higher and higher frequencies, the lines crowded closer and closer together, as you saw in the diagram. Here, then, are the first four numbers that Ångström published." She pointed again to the text of her article:

$$H_\alpha = 6562.10$$

$$H_\beta = 4860.74$$

$$H_\gamma = 4340.10$$

$$H_\delta = 4101.20$$

"When he took the ratio of the first two numbers, Balmer noticed something very interesting."

6562.10/4860.74 = 1.350020779

"There were two zeros right after the first two decimal digits. What if the 1.35 was important, while the . . . 0020779 . . . part was due to the expected observational errors by Ångström? So Balmer rewrote the 1.35 part as a fraction:

135/100 or, in reduced form 27/20

She had lost me, so I interrupted to ask, "What ever possessed Balmer to take the ratio of the wavelength numbers?"

She looked at me dubiously, as though it were a stupid question. "The first thing you would want to do is eliminate any obscuring common factors, especially ones that might be irrational or complicated in other ways. For example, the irrational number pi is 3.14159 to five decimal places. Now 2 times pi is approximately 6.28318, while 3 times pi is 9.42477. If you take the ratio of these two numbers, you will get 2/3, in effect, a rational number.

"Now when Balmer took other ratios among the four numbers, such as H_α to H_γ, the same thing happened. He named the common factor *b* and called it 'the fundamental number of hydrogen.' After eliminating *b*, however, what kinds of number remained? Were they simple integers or perhaps ratios of integers? After all, if you take a ratio of ratios, you still end up with a ratio."

I had to think for a little while to realize that she was, of course, right. For example, that ratio of 3/4 to 8/5 is simply the fraction.

$$\frac{3/4}{8/5} \text{ which equals } \frac{3 \times 5}{8 \times 4} \text{ which is } \frac{15}{24}, \text{ a ratio again.}$$

Just then the music from San Marco stopped, and soon after, people flooded out of the church to mill about on the piazza. It was intermission.

"As I was saying, Balmer probably tried to find the integers lurking in these numbers by first assuming that each of the Ångström numbers was of the form *bm*, where *b* is the fundamental number of hydrogen and *m* is an integer. But he got nowhere with this approach. Evidently, he was dealing with ratios of ratios. In the process of discovering what these ratios were, he may well have proceeded by assuming that each of the Ångström measurements had the same general form:

$$b(n/d),$$

where *b* is the fundamental number and *n*/*d* is the ratio lurking in the measurement. Here, *n* stands for numerator and *d* stands for denominator.

"Now we can use algebra as a kind of microscope to see what happens in the general situation. Here are two such numbers, with their components subscripted so as to distinguish them. I have formed their ratio:

$$\frac{b(n_1/d_1)}{b(n_2/d_2)}$$

"Algebra tells us that when we take the ratio of such numbers, the *b* factors cancel out, and we are left with a ratio of ratios, which boils down, in this general form, to yet another ratio of integers. Notice that each integer in the new ratio is a product of two others.

$$\frac{n_1 d_2}{d_1 n_2}$$

"Armed with this kind of notation, Balmer would have quickly turned the whole problem into a set of equations that could be solved by routine methods. For example, remember that when he took the ratio of the first two hydrogen numbers, H_α and H_β, he got 27/20? All he had to do now was to set this ratio equal to the general form I just showed you. The rest would then fall out by algebra:

$$\frac{n_1 d_2}{d_1 n_2} = \frac{27}{20}$$

"He also used two other equations, one for each of the other possible ratios:

$$\frac{n_1 d_3}{d_1 n_3} = \frac{189}{125}$$

and

$$\frac{n_1 d_4}{d_1 n_4} = \frac{72}{45}$$

"In the end, these three equations were all Balmer had to go on. As things turned out, they were all he needed. Now there were not three equations, but six. For, in each of the three equations, he set numerators equal to numerators and denominators equal to denominators.

$$n_1 d_2 = 27$$

$$d_1 n_2 = 20$$

$$n_1 d_3 = 189$$

$$d_1 n_3 = 125$$

$$n_1 d_4 = 72$$

$$d_1 n_4 = 45$$

"Balmer now had six equations in eight unknowns. It was a mathematical fact, known even to the young ladies to whom he taught mathematics, that a system with more variables than equations usually has more than one solution. Once the equations were set up like this, it would have taken him no more than an hour to solve them. It's really quite amazing how quickly the solutions fall out. Watch this!"

She pointed at the first equation, $n_1 d_2 = 27$. This meant that whatever value the integers n_1 and d_2 had, the product of those values must be 27. There were essentially two possibilities. Either $n_1 = 9$ and $d_2 = 3$ or the other way around, $n_1 = 3$ and $d_2 = 9$. When the first set of values was substituted into the equations, its effect rippled through, leading to definite values for some of the variables, while leaving others still unknown. If you substitute 3 for n_1 and 9 for d_2, wherever these variables appear in the previous set of six equations, you get the following, somewhat simpler set:

$$\underline{n_1 = 9,\ d_2 = 3}$$

$$d_1 n_2 = 20$$

$$d_3 = 21$$

$$d_1 n_3 = 125$$

$$d_4 = 8$$

$$d_1 n_4 = 45$$

The three unchanged equations all contained d_1 as a factor, and together they implied that whatever this number might be, it would have to divide the three integers 20, 125, and 45 evenly. The only number to do this is 5, so the equations forced the following values, all of them stemming from the original assumption about n_1 and d_2:

$$\underline{n_1 = 9,\ d_2 = 3,\ d_1 = 5}$$

$$n_2 = 4$$

$$d_3 = 21$$

$$n_3 = 25$$

$$d_4 = 8$$

$$n_4 = 9$$

Canzoni continued, "Now that he had values for all the integers participating in the wavelength ratios, he could go back and substitute them into the formula $b(n/d)$, which he had assumed to hold for the wavelength numbers. The results must have pleased him very much, for his first thought was surely to extract the fundamental number of hydrogen. Take the case of the first hydrogen line at wavelength 6562.10.

"He could now write the equation

$$b(n/d) = 6562.10,$$

substitute the values $n = 9$ and $d = 5$ to obtain

$$b(9/5) = 6562.10,$$

then to solve for b, which again, is simple algebra:

$$b = 3645.6$$

"When he tried this with the other ratios, such as n_2/d_2, solving for the fundamental number in each case, he got values that were very similar. Here they are, all together: 3645.6, 3645.5, 3645.7, 3645.5.

"It was truly remarkable. The fact that each set of n and d values produced essentially the same value for the fundamental number of hydrogen meant that Balmer could then use this value in the original formulas and reproduce the wavelengths:

$$3645.6(9/5) = 6562.08,$$

which agreed with the original wavelength, 6562.10, to better than one part in a hundred thousand. He had found his integers!"

I felt somewhat suspicious, as though Canzoni were pulling the wool over my eyes, or, worse yet, over her own. The results were almost too good to be true. I immediately remembered there had been two solutions. Where had the other values for n_1 and d_2, with the 9 and 3 reversed, led Balmer? She replied that they had led to essentially the same solution.

"It strikes me as ironic," I said. "Had someone proposed this little problem to a working mathematician without saying where the data came from, he or she would probably find it somewhat boring."

"Absolutely. By itself, it is not a particularly interesting problem, and I am sure that, had it been presented to Balmer in any other context, he would not have given it a second look. Balmer had certainly not discovered any new mathematics. He had merely applied algebra—and nothing very sophisticated, at that. But here was a message from the cosmos in the form of four little numbers. Balmer had deciphered the message, at least to the extent of discovering four integer ratios hidden in their hearts. But that was the easy part. What he did next was more interesting, mathematically.

"He looked closely at the series of ratios he had obtained for the four hydrogen wavelengths: 9/5, 4/3, 25/21, and 9/8. He noticed that the numerators were all squares, while the denominators were either 1 or 4 less than the numerator. The ratios, in short, formed an interesting-looking series. To mathematicians, this sort of suggestive structure is like waving a red flag in front of a bull. Off he runs! The mathematician wants a formula that will generate these ratios, ad infinitum if necessary. In such formulas, you have an integer variable, say m, that acts like a counter. The variable m takes on the values 1, 2, 3, . . . and so on, while the formula churns out the Balmer ratios such as 9/5, 4/3, and so on.

"He toyed with formulas such as

$$\frac{m^2}{m^2 - 1} \quad \text{and} \quad \frac{m^2}{m^2 - 4}$$

"When he substituted $m = 2, 3, 4$, and so on into these formulas, he got back all the ratios he had derived earlier, and more besides, as I will show in a moment. His real genius, as such, came from a leap of faith—you could call it Pythagorean faith—and proposed the following general formula, one that contained the previous two.

$$\frac{m^2}{m^2 - n^2}$$

"Each value of n led to a different series. When $n = 1$, the formula boiled down to the first formula I wrote earlier. It produced the numbers 4/3, 9/8, 16/15, 25/24, and so on, when you substituted $m = 2, 3, 4, 5$, and so on. When $n = 2$, the general formula became identical to the second formula I wrote earlier. Here, with $m = 3, 4, 5, 6$, etc., the formula produced 9/5, 16/12, 25/12, and so on. You will find in these two series each of the four ratios that Balmer discovered in the cosmic data from Ångström. Then came the clincher. Not only did every new measurement that Ångström sent to Balmer fit his formulas, but Balmer also predicted new hydrogen lines as, for example, in this quote from a paper published in 1885:

> From the formula we obtained for a fifth hydrogen line 49/45.3645.6 = 3969.65.10^{-7} mm. I knew nothing of such a fifth line, which must lie within the visible part of the spectrum just before H_1 (which according to Ångström has a wavelength 3968.1); and I had to assume that either the temperature relations were not favorable for the emission of this line or that the formula was generally not applicable.

"However, that line was found and many more besides. Today the series predicted by Balmer's formula have all materialized. That is, more refined spectrographic techniques have revealed a great many new lines of the hydrogen spectrum. They are known as the Lyman series ($n = 1$), the Paschen series ($n = 2$), the Brackett series ($n = 3$), and the Pfund series ($n = 4$). In short, all of the physically possible lines predicted by Balmer's formula turned out to occur in natural hydrogen—and no others."

At this time, a bell from inside San Marco sent the music lovers flooding back through the doorway and past us, into the church. Some looked quizzically at the two of us talking animatedly on the bench—two music critics, no doubt, deep in a discussion of the development of polyphony. The pause gave me time to reflect. I wanted to leave no stone unturned.

"In your view, was there any other solution or any other formula that could have arisen from these four numbers, not to mention the ones that came later?" I asked.

"I cannot even conceive of the possibility," whispered Canzoni, watching a lone gondola making its way toward the harbor mouth. "You would sooner see that boat take wings and fly to the moon. You could labor a lifetime and not find another formula. You see, there is only one, and Balmer found it.

"Now the significance of this formula was not really understood for many years. In the very year that Balmer's paper appeared, a boy was born to a family named Bohr in Denmark. It was Niels Bohr who would finally explain Balmer's formula when he investigated the new quantum model of the hydrogen atom. The lines corresponded to the energy levels that such an atom might have. And each level produced its own characteristic radiation of a specific wavelength, the numbers measured by Ångström.

"The new quantum theory, as developed by Bohr and others, had at its very foundation the idea that energy was ultimately not continuous at all, but discrete. The quantum numbers that indicated these states were all integers or half integers. Balmer, unfortunately, did not live to see the new theory, to see Pythagoras's integer cosmos reborn in this manner."

She turned to me.

"I am sorry that we were late for the performance. It is my fault. I did not watch the time very closely. I will take you across the Grand Canal to your hotel."

We walked slowly over the massive paving stones. It was late in the evening now, and the smell of the city had changed, it seemed—it was now more complex, a potpourri of ancient stone, food, motor oil, rotting wood, pollution, and God knows what else. I took a deep breath, inhaling Venice like a tonic, and asked Canzoni what she had thought of my three questions. What, in particular, did the story of Balmer illustrate?

"Before I answer, I must say that your questions are too weak. It is like the parable of the blind men and the elephant. One of the blind men asks, 'Why does this animal have meter-long teeth?' Another asks, 'Why does this animal have skin like a wrinkled blanket?' Someone should ask, 'What is this elephant doing here?'

"In the same way if you ask, 'Why does mathematics seem to be discovered?' or 'Why does mathematics keep showing up in the physical world?' you are missing the point of it all. The elephant has been here all along, but we are blind."

"What elephant?" I interjected. I was losing track of her point.

Canzoni suddenly laughed. "All I can tell you is that it's an invisible elephant—invisible to our ordinary senses, that is. To those who ponder the cosmos and its connection to mathematics, the elephant can be sensed.

"Balmer's discovery was not a mathematical discovery, as I have already pointed out. So his findings do not address the independent existence of mathematics, at least not directly. But they squarely address the presence of mathematics in the cosmos.

"Balmer's formula was discovered in the middle of four numbers that came to us from distant stars. Those numbers, undulating through space in the form of four precise wavelengths, contained a message of sorts, one that was uniquely decipherable within the framework of mathematics. In fact, we can even prove, by the methods of information theory, that there was room within those 20-odd digits for just one message of the size that Balmer discovered. The message was a formula for the energy levels of the hydrogen atom.

"Alien beings would find precisely the same message in those numbers, albeit they might express the formula very differently. Balmer's discovery provides a striking example of the mathematical patterns that reside in nearly every aspect of physical reality when you examine it closely enough. I have a thesis, which I will explore with you tomorrow. It says that those patterns are there because something within the cosmos satisfies axioms within the holos. It explains the invisible elephant.

"In the meantime, I have a question of my own. Why do some people resist so stubbornly the idea of an independent existence to mathematics? Before one gets more than a sentence out, they begin to squirm and wriggle, to gaze at the ceiling. Something, I think, is bothering them, as though I were violating their freedom. Other people find the idea of underlying structure wholly acceptable. 'Why not?' they wonder.

"Let us be frank. Even the philosophers who find reason to doubt the existence of objective reality are quite happy to act as if one existed. How could they do otherwise and continue to live? Without calling them hypocrites, I observe that they behave as if they believe in an objective reality. Physical reality has enough stability for them to plan, imagine, and remember without making major mistakes most of the time.

"And if we accept the existence of an objective reality and if we agree that it contains lawful regularities, as discovered by physics and other sciences, and if we also accept that mathematics has an independent existence of a quite peculiar kind, then what, I ask, is the simplest possible explanation for this state of affairs? The holos controls the cosmos because the cosmos has no choice in the matter. If a particular physical system within the cosmos obeys certain axioms, in effect, how can it fail to obey, as well, every single theorem that holds true for those axioms?"

We had come to the midpoint of a bridge across the Grand Canal. Canzoni stopped to gaze into the slow, murky current.

"Sometimes, you know, I, too, rebel against this state of affairs. Sometimes, when I contemplate the holos and the cosmos together, I find it the most terrifying thought in the world!"

I was astonished. "Why?" I asked.

"It is hard to put into words. You see, the independent existence of the holos may in some sense explain the cosmos, but there is still the existence of the holos to explain, and that is utterly beyond my powers. It is both reasonable an unreasonable. All I can tell you is that we humans can only grasp the holos through our minds. What if its ultimate existence is also a mental phenomenon of some kind? Not our minds, but another."

She said nothing more. We walked quietly to my hotel. In front, I asked her how she would get home. She answered that she did not live far away.

"Tomorrow, as I said earlier, I will show you something interesting." With that, she turned and left abruptly.

My room was luxurious, with marble shower, telephone in the bathroom, and a basket of fruit on the table. The cost, according to the night clerk, came to 30,550 lira, which, I hoped, would not bankrupt me.

I stood on the balcony of my room and looked out over the Grand Canal. If Canzoni was right about the message in the hydrogen lines, that it was uniquely decipherable, it made a persuasive case for the intimate linkage between mathematical patterns and physical reality. The holos and the cosmos were linked somehow. Was it as simple as she had claimed? Do physical systems truly obey axioms? If so, it would be as Canzoni had said. How could such systems avoid obeying all the theorems that flowed from those axioms? What would Canzoni tell me tomorrow?

I lay down on the bed, intending only to rest for a moment, but I fell into a deep sleep almost instantly. I awoke to the sound of thunder crashing all around me and the memory of a disturbing dream. It was 2 A.M. Amid flashes of lightning, details came back to me. I had drowned in one of the canals, sinking into a cold darkness that weighed upon me. I don't remember dying, but something in the dream said that I had. I witnessed a sudden flash of light and heard the roar of manifestation. The universe was born in a thought that had no thinker. At least, it wasn't me.

CHAPTER 6

The Ultimate Reality

—⁓—

In the Historia della Scienza building, just off the Grand Canal, a series of cunning balconies made generous alcoves along the length of its third story. It was in one of these (equipped with its own chalkboard!) that Canzoni and I sat the following morning.

Apart from the disturbing dream, I seemed to have slept very well. The thunderstorm was long past. Bright sunlight shone everywhere, and faint breezes stirred the scents of Venice, wafting them up from the walks, the canals, the buildings themselves. I felt alert, anticipating the surprise she claimed to have in store for me. Would I see the elephant vanish?

Canzoni seemed as fresh as the morning, as if her link to Venice were directly physical. "To be frank, I must express my gratitude to you. Your visit has energized me and that single word, the one from Pygonopolis, has by itself enlarged my view. Sometimes, to name a thing produces the most extraordinary results. In my case, it has given me the courage to dust off a theory that I have been keeping in my, my *cabinetta,* so to speak, for many years. I will say more about it later. In the meantime, the word *holos* is now part of my working vocabulary."

She then asked me about my views of the holos and what features it had that might persuade me that it had an independent existence.

"It seems to me," I said, "that the question of independent existence might be cast in geographical terms. The holos, whatever we take it to be, has a sort of structure. I mean if two mathematicians start from the same axioms, they will, eventually, discover the same things very frequently, like two explorers wandering over the same island. In their logbooks, they might both make entries. Explorer A writes, 'Due south of the coastal mountain shaped like a loaf, I found a deep embayment with a fine beach stretching its entire extent.' Explorer B meanwhile writes, 'I wandered through dense jungle due east until I came to a vast beach that stretched for more than a mile. It lines a bay guarded at its north end by a sentinel of rock, a huge hill in the shape of a termite mound.' If both explorers drew maps, however crude, it would be clear that, among other features, they had both discovered the same beach. And no one would think it the slightest bit remarkable."

"Bravo. That was nicely expressed," laughed Canzoni. "Most romantic. Still, it must be said that we have yet to carry out a controlled experiment in which two mathematicians do precisely that. We must take our data from actual historical events. Surely you are aware how frequently different mathematicians have stumbled on precisely the same theorem. They do not even have to live at the same time or belong to the same culture.

"You know perhaps the classic case of independent discovery. In what would later be regarded as a culminating development of early European mathematics, Newton and Leibniz both discovered the calculus. The more closely you look at their work, the less surprised you are at the phenomenon. In the early eighteenth century, natural scientists were consolidating the laws of motion, as described by Galileo and others. They sought a technique for analyzing motion but were defeated by the fact that the new Copernican system, in its Keplerian formulation, demanded that they deal with continuously varying quantities. A body projected upward did not move uniformly, for example, but slowed under the influence of gravity, eventually stopping, then falling earthward with increasing velocity.

"Both men had access to the newly discovered analytic geometry of René Descartes. One could see at a glance what a physical system was doing. Here, for example, is the curve followed by a projectile in a gravitational field. Let us say it is a stone thrown straight up at 20 meters per second."

She went to the blackboard and drew a figure like the following:

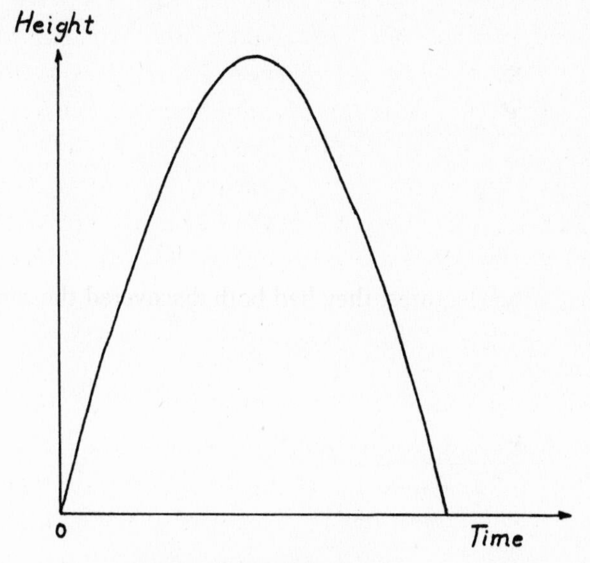

Height of Stone Versus Time

"Here is a curve that portrays the ascent and descent of the stone. The horizontal axis represents time, the vertical one the height of the stone above the ground. For each moment, from the instant the object is thrown to the time it lands again on the ground, the stone has a certain definite distance above the ground, and a certain, definite velocity with respect to the ground.

"I could paraphrase the experience that both Newton and Leibniz had with such a curve by saying that both recognized that the vertical velocity of the stone at any time was closely connected to the slope of the curve at that time."

She drew a right-angle triangle over the figure. The base of the triangle represented a certain elapsed time, and its vertical side represented the distance upward traveled by the stone over that time. The slope of the hypotenuse of the triangle represented the velocity of the stone over the time interval. The slope was simply the ratio of the vertical side to the horizontal side, the tangent of the angle at the base, as al-Flayli might have said.

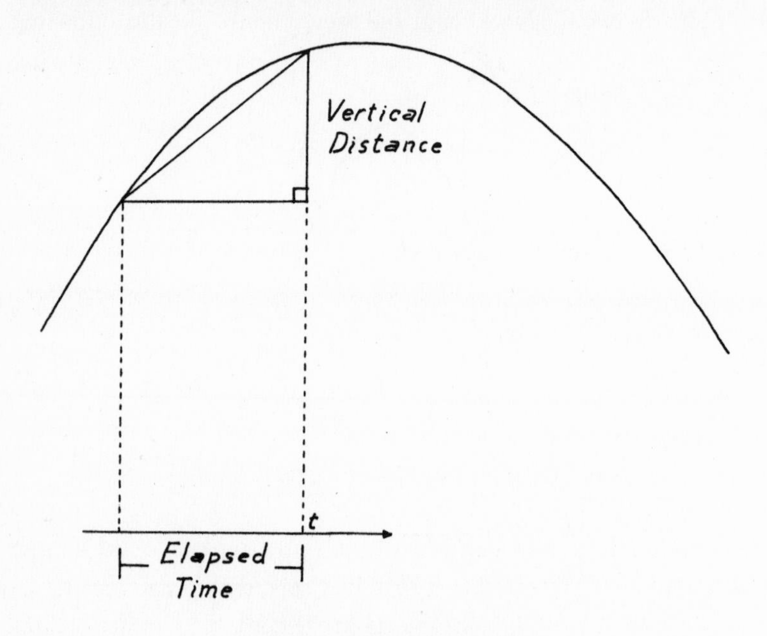

Velocity Triangle of the Stone

Canzoni continued, "Both scientists, working now in their capacity as mathematicians, could see that the hypotenuse of this triangle was only an approximation, a kind of average velocity of the stone over the time interval represented by the base of the triangle. But both could also see that as they made their triangles smaller and smaller, continually shrinking the time base, the slope converged (or appeared to converge) on a specific value, which could only be the *instantaneous velocity*—that is, the velocity at the moment in question. See, here I am shrinking the triangle, and here I am watching the hypotenuse get closer and closer—to what?"

Canzoni had undergone a remarkable transformation while explaining the diagram. Her movements became forceful and aggressive. She shrank the triangle first by mime, as though it were a physical operation. Then she drew a succession of smaller and smaller right-angle triangles. With successive triangles, she swayed, shifting from hip to hip, as though dancing.

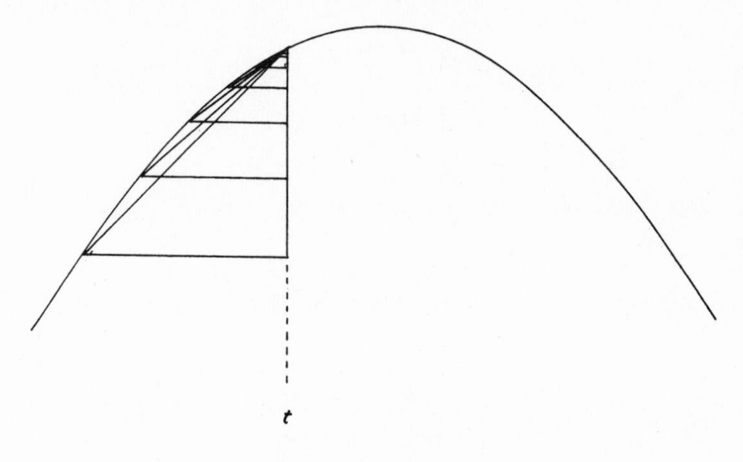

A Succession of Triangles

It occurred to me, just as she turned to watch my reaction, that she sometimes lectured precisely like this, that she harbored the soul of an actor expert in the communication of abstract ideas through bodily motions. Some philosophers have claimed that mathematics is the visceral science, almost tactile. Mathematicians wrestle with problems like everybody else, but what wrestling! Canzoni wrestled this clarity of perception out of nowhere, turned, and smiled.

It was my turn to say, "Bravo!"

"They could both see," she continued, "for every curve they could conceive, how the shrinking triangle would eventually vanish to nothing, but how the slope of its hypotenuse would somehow remain, still present at the final vanishing point as a very special line, known even in antiquity. Today we call it the tangent line. It just touches the curve at the tangent point, where the velocity is being investigated."

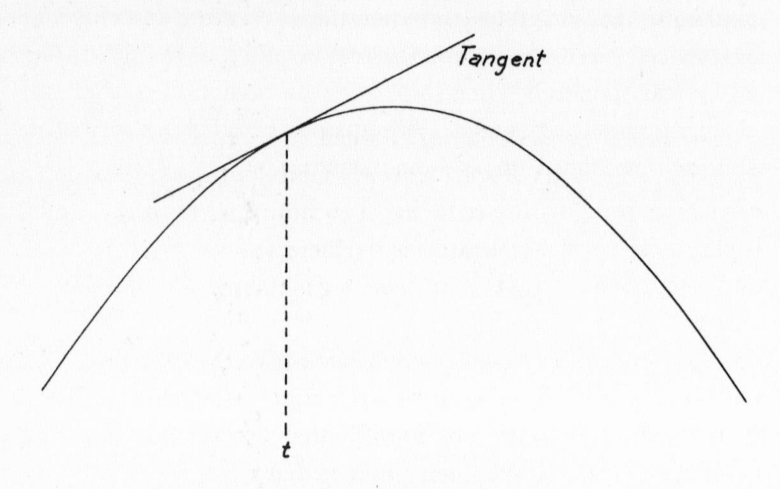

Velocity = Slope of the Tangent

"As far as actual tools by which to analyze this peculiar situation, Newton and Leibniz were equally at a loss. I mean tools for proving things, tools to produce a state of certainty. It was one thing to see that the slope of the ultimate hypotenuse, the one that vanishes, is the same as the slope of the tangent to the curve, but it was quite another to prove it. How could they talk sensibly about the slope of a line that had vanished to nothing? In their own particular ways, Newton and Leibniz simply assumed what was visually obvious. The slope of the tangent to the curve at the point in question was, in fact, the instantaneous velocity at that point.

"I should add that in the early eighteenth century, curves such as the one I just drew had algebraic formulas, thanks to René Descartes, among others. For example, the equation of the stone's vertical position would have been written in rather similar fashion to the way we write it today."

She scribbled the following equation on the board, explaining that y was the stone's vertical position, as measured from the ground, and that 20 was the initial upward velocity with which the stone was thrown. Time t started at 0, the precise instant the stone was launched. For each value of t, the corresponding value of y would give the vertical height of the stone.

$$y = 20t - 4.9t^2$$

"What is the negative term?" I asked.

"That is the effect of gravity, always subtracting away the stone's vertical velocity, finally reducing it to nothing, when the stone begins to fall again, accelerating at 9.8 meters per second every second. The general term is $\frac{1}{2} gt^2$, where g is the acceleration due to gravity.

"As I was saying, Leibniz and Newton both discovered that an equation of position such as the one before us could be reduced to an equation of motion by multiplying each coefficient by its accompanying power of t, then reducing that power by 1. So it was that

$$20t^1 - 4.9t^2$$

became

$$1 \times 20t^0 - 2 \times 4.9t^1$$

or simply

$$20 - 9.8t$$

"The equation of motion in which y, the vertical position, was replaced by the vertical velocity, now gives the upward velocity of the stone at each possible instant. The process of going in this manner from a formula for the position of the stone to a formula for its velocity was the key component of what we know today as the differential calculus.

"Newton, who made his discovery a couple of decades before Leibniz, kept the method secret. He called the new mathematics *fluxions* and wrote the operation we have just described as $\dot{y}$, where

$$\dot{y} = 20 - 9.8t$$

Leibniz, on the other hand, called it the differential calculus and wrote the velocity in his own "differential" notation as

$$dy/dt = 20 - 9.8t$$

"As you know, Leibniz's notation and terminology were the ones to survive, not Newton's. They had, of course, both discovered exactly the same method for dealing with motion. Both realized that the new mathematics involved calculating not only differentials but integrals as well. The latter operation was essentially just the reverse of the differential. For example, in the context of this example, the integral of

$$20 - 9.8t$$

was simply

$$20t - 4.9t^2$$

"If the differential meant determining the slope of a tangent to a curve, the integral turned out to involve determining the area under a curve.

"The new calculus came to be applied to just about every conceivable type of motion or change in the physical world. Often, mathematicians or physicists would begin with an equation that involved differentials, moving by integration to an actual formula of position, such as the one we started with here. Such equations, called *differential equations*, have dominated physics ever since. They appear in Schrödinger's equation of the hydrogen atom and in Einstein's theory of general relativity.

"It was only near the end of the eighteenth century that mathematicians, engaging in a great fit of housecleaning, attempted to put the new calculus on a more rigorous footing. The French mathematician Augustin Cauchy did much of this work, establishing that when one shrank the right-angle triangle, as I showed you before, the slope of the ultimate hypotenuse, the one that vanished, was in a perfectly rigorous sense equal to the slope of the tangent at that point. The calculus was now secure.

"Before that time, the followers of Newton or Leibniz used the new technique with a kind of gay abandon. There were heated arguments about which, uh, maestro had priority in the great discovery. But there was little discussion about the logical fundamentals of the differential calculus.

"So it was that two great minds had entered the holos like explorers. They came from different directions, sailed in different ships and at different times, but came upon the same new continent, finding there the same geographic features. Their logbooks contain different entries and expressed their discovery in different language, but anyone could see they were describing the same thing. How else could there have been such a dispute about priority?"

It was time, I thought, to get back to the holos of which Canzoni had become so enamored. With that in mind, I asked, "So this illustrates the real existence of the holos?"

"If you mean *real* in the very special sense we have discussed, yes. Does this case of independent discovery not illustrate the independent existence of the holos? And do not literally hundreds of cases of independent discovery not illustrate it further? There have been far, far more of these than one can account for by some kind of chance, even by cultural influence. I might remind you that the number of possible formulas or expressions, counting only ones that are completely inequivalent, is still infinite. If you think the holos has no independent existence, then even a single case is a miracle."

"And what about the cosmic connection?" I asked. I was being perhaps a bit silly, but she knew what I meant.

"The equation for the stone's position amounts to an accurate description of its behavior at all points of its flight. Some people object that the stone will encounter air resistance and so the path shown will not be the true one. But this is a quibble in the sense that we also have a complete theory of air resistance, and a single term suffices to make the equation even more accurate. It is no accident that the equation is so accurate. It embodies two of Newton's laws. There is first the upward momentum of the stone, which, as a body in motion, tends to remain in that state of motion unless impelled by some external force—in this case, gravity. As for gravity, that was perhaps Newton's greatest achievement. The effect of gravity on any body that was free to move was to accelerate it. Momentum and gravity are simply facts. They can both be measured in the laboratory, any time, any place, and the result will always be the same. They are part of the underlying structure of the cosmos, if you like. *Why* they should be so is another question entirely."

She drew a long breath. "There is another example of simultaneous discovery, not in mathematics this time, but in astronomy. But this example is really about the intimate connection between the cosmos and the holos. It concerns the accuracy of the new law of universal gravitation that Newton had discovered. It involves two mathematicians, both young and both applying the celestial mechanics of Newton."

"You say it was not a case of simultaneous mathematical discovery, yet two mathematicians were involved?"

"They applied mathematics to a celestial phenomenon, discovered exactly the same possibility, and made exactly the same prediction. Here is the story."

She returned to her chair and took a sip of coffee.

"In 1781, the astronomer William Herschel had discovered a new planet that he called Uranus. Other astronomers pointed their telescopes at the new body and plotted an orbit for it according to the new celestial mechanics. At first, Uranus followed its plotted orbit dutifully, but, after many years, certain discrepancies began to creep in. The planet was moving more slowly in its orbit than it should be according to Newton's theory. What was wrong? Was Newton's theory, so beautifully worked out, now to be discarded, along with the epicycles of old? The British Astronomer Royal, George Airy, worried that perhaps the Newtonian law was not universal after all, but weakened more quickly with distance than was thought, perhaps disappearing altogether at great enough distances. But he could not guess how this might be.

"In 1843, John Couch Adams, newly graduated from Cambridge University, set to work in earnest on a most laborious problem. If the irregularity in the motion of Uranus was due to yet another, unknown planet orbiting farther away from the sun, it was theoretically possible to discover the orbit and position of that planet, using the same laws of universal gravitation that had made it possible to predict the motion of Uranus and the other planets in the first place. By 1845, Adams succeeded in making a prediction of the existence of a new planet. He sent a letter to Airy, telling him where to look in the sky, but Airy was away in France at the time. When Airy returned, he was instantly curious about Adams's result and wrote him a letter in turn, inquiring further into the discovery. Adams went to see Airy later but missed him again, and thereafter, perhaps feeling rejected, he did not bother to reopen the correspondence.

"Meanwhile, in France, the only slightly older Urban Jean Leverrier was set loose on the same problem that Adams had solved earlier. Not knowing of Adams's claims about the existence of a new planet, Leverrier proceeded to calculate, with much the same labors, the orbit and mass of the new planet, ending with a position in which astronomers might seek the new heavenly body. This position differed from the one given by Adams by less than a degree. Leverrier informed not only Airy of his prediction, but astronomers in Berlin as well. The British, being somewhat slow off the mark, were beaten by the Germans, who confirmed the existence of a new planet, right where Leverrier had said it would be.

"A few years later, both Leverrier and Adams were awarded gold medals for the discovery of Neptune by Sir William Herschel, who, on that occasion, said . . . "

She broke off in midsentence, got up, and left the balcony, returning a minute later. "Here is a copy for you. This is part of Herschel's speech on the occasion of the awards." I read it, a photocopied page from a book on the history of science:

> The history of this grand discovery is that of thought in one of its highest manifestations, of science in one of its most refined applications. So viewed, it offers a deeper interest than any personal question. In proportion to the importance of the step, it is surely interesting to know that more than one mathematician has been

> found capable of taking it. The fact, thus stated, becomes, so to speak, a measure of the maturity of our science; nor can I conceive anything better calculated to impress the general mind with a respect for the mass of accumulated facts, laws, and methods, as they exist at present, and the reality and efficiency of the forms into which they have been moulded, than such a circumstance. We need some reminder of this kind in England, where a want of faith in the higher theories is still to a certain degree our besetting weakness.

"It was a culminating event," Canzoni broke in.

"What sort of culminating event?" I asked.

"As Herschel said, it marked the maturity of the Newtonian theory of gravitation. Adams and Leverrier, working completely independently, used the Newtonian theory to discover a new planet. Not only did the independence establish that the calculations were correct, but also the accuracy of the predictions actually confirmed the Newtonian theory and allayed Airy's fear that gravitation did not work so far out. In any event, the holos not only told them there was a planet they didn't know about, but it also told them where it would be."

"What a pity," I sighed. "Now that the Newtonian universe has been supplanted by the Einsteinian universe—"

"Nothing of the kind," she interjected somewhat archly. Do you remember the invisible elephant from yesterday?"

I nodded my head and kept a serious expression on my face.

"If the Einsteinian universe, the one governed by relativity, is the leg of the elephant, then the Newtonian universe is the foot, in a sense. General relativity, as the name implies, is general. It describes the dynamics of objects traveling at relatively slow velocities, down at the foot, as well as objects traveling at much higher speeds, the rest of the leg, up to the speed of light itself, the absolute limit to all velocities in the cosmos, about 300 million meters per second.

"All the speeds we deal with in daily life, even the speeds of fast-moving spacecraft, may be found down in the foot of the elephant, so to speak. At such speeds, Newton and Einstein agree. Let me give you an example. According to relativity theory, the clocks aboard a fast-moving spaceship will appear to be running slowly when compared to clocks back on Earth. How much more slowly? It depends on the speed. The more rapidly the ship goes, the more slowly its clock will appear to run. The actual correction factor is very easy to compute.

$$= \frac{1}{\sqrt{1 - v^2/c^2}}$$

The clock aboard a spacecraft traveling at velocity *v* will appear to Earth-bound observers to be running slowly by this factor. Suppose the spacecraft travels at, say, 3 million meters per second. That's much, much faster than any spaceship we have built yet. What is the Einsteinian correction in this case?" She wrote the following calculation on the board:

$$\frac{1}{\sqrt{1 - (3/300)^2}}$$

$$= \frac{1}{\sqrt{1 - (0.01)^2}}$$

$$= \frac{1}{\sqrt{1 - 0.0001}}$$

$$= \frac{1}{\sqrt{0.9999}}$$

$$= 1.00005$$

"As you can see, the correction factor is so close to 1 that it hardly makes any noticeable difference. The clock on board the spaceship will be about 30 seconds slow after a week of travel at this speed.

"Oh, yes, I nearly forgot. The theorem of Pythagoras is there in the formulas that we use in special relativity. It's the square-root factor in the bottom of the fraction. I won't go into it, but that factor comes from Pythagoras!"

She reached for her coffee, which she had by her chair, and took a sip, glancing briefly at a pigeon that flew to the railing of our balcony, like a signal.

"I think it is time for the elephant to vanish." She spoke as if she were a magician, about to go onstage. She got up abruptly and erased the diagram of the falling stone and its formulas from the board. Then she wrote a sort of title:

MATTER → ENERGY → INFORMATION

"You could summarize my ideas about the holos this way," said Canzoni. She stood for a full minute staring at what she had just written. "In 1805, John Dalton published his first paper on the atomic theory. It was surely not the first such publication. The Greeks had toyed with the idea of atoms, as had the Romans. Read Lucretius, and you will find a surprisingly modern account of the idea. But with Dalton, for the first time, we learned of atoms. They were hard little particles that composed all matter. Dalton imagined that if these particles could be magnified, his atoms would resemble spherical shot or pellets. Dalton's atoms also combined in certain specific ways to produce various compounds, which formed the beginning of chemistry, but I don't want to get sidetracked. The point is to think about atoms as little pieces of shot.

Daltonian Atom

"Now the Daltonian atom, even though it took a few decades to become widely accepted, represented a revolution in thinking no less important than the Copernican revolution. Yet no one speaks of a paradigm shift in relation to it. In fact, it was more important because it was about things you could touch and feel, not to mention yourself, and not about distant bodies in space. The new atomic theory addressed something that all people could hold in their hands and wonder about.

"Let us say that you are knocking on a door, seeking admittance. Your knuckles strike the wood of the door. Knock-knock. The door is very hard. In a sense, this knocking on a door, the feeling of that hardness, that is what most people think of when they think of reality. Now, when thinking people learned about the Daltonian atom, that sense of knock-knock reality itself took a knock. The knuckle that was composed of hard little atoms struck the door that was composed of hard little atoms. If, in the end, this took some getting used to, it was not so bad because that hard, knock-knock quality had been transferred to the hard little spheres that composed matter. Matter, after all, was matter."

I wondered what she meant by *transferred* and asked her.

"I mean that when they felt a little dizzy from thinking about something solid like a door being made up of untold billions of little spheres called atoms, they could at least think of those spheres, as Dalton did, as hard and durable in the same way that they would originally have thought of the door.

"Unfortunately, by the end of the nineteenth century, our view of matter had changed radically once again. Atoms turned out not to be hard little spheres, but to have a structure. Still pretty much spherical, they now consisted of a tiny nucleus at the center, with electrons whizzing around the outside. In between the nucleus and the electrons, more than 99 percent of an atom was empty space.

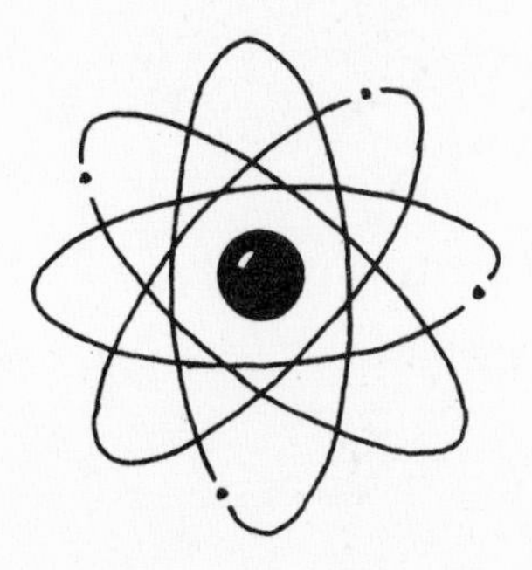

The Rutherford Atom

"Now people had to deal with a view of matter that was mostly empty space. Nevertheless, both the electrons that circled atoms and the protons and neutrons that composed their nuclei could still be viewed as the ultimate repositories of knock-knock reality, but the game was getting a little difficult to sustain. It was becoming increasingly difficult to hold to the intellectual life raft of tangible reality. Then came the next big shock." She paused.

"What shock?" I asked.

"With the dawn of the twentieth century, the physicist Einstein showed that matter and energy were equivalent. A small amount of matter, m, contained enormous amounts of energy, equivalent to mc^2, as everyone knows. The speed of light, c, is a very large number and when you square it, the number becomes much larger. Every atom, every particle of every atom—whether electron, neutron, or proton—contained energy, consisted of energy. The point is that energy not only inheres in the matter, it *is* the matter."

I was becoming a little uncomfortable with the drift of things. "I thought that energy and matter were merely interconvertible," I remarked.

"True, but the energy is always there, in the heart of every particle, waiting to manifest, if you like. Energy is the ultimate constituent of all matter. That is what I mean. Indeed, most twentieth-century physics is about energy. It either resides for a time in some particle or bursts forth as a wave. All the fundamental constituents of the atom were now viewed essentially as arrangements of energy that created force fields.

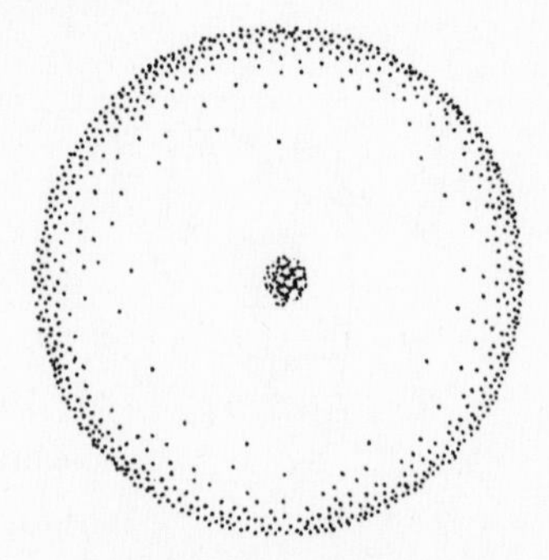

The Atom-as-Energy

"Those who pondered the ultimate nature of things now felt more adrift than ever. The knock-knock reality had disappeared entirely, to be replaced by energy. The fist was a huge arrangement of energy that approached, interacted with, then rebounded from the door, an even larger arrangement of energy.

"Then came Bohr and the group of physicists who worked with him, known as the Copenhagen School. They developed a view of atomic reality so bizarre that even Einstein refused to believe it, in spite of the steadily accumulating evidence that supported it. The energy of the nucleus, of the electrons, of waves in space, of all manifestations of reality, was parceled into tiny packets called *quanta*. The energy of the lone electron orbiting a hydrogen atom could not take on just any old value, but only multiples of a fundamental value. This explained the behavior of excited hydrogen atoms. They emitted energy or light only at specific wavelengths that corresponded to those quantum levels. The lowest quantum level corresponded to the first line of the hydrogen spectrum that we discussed yesterday. The next quantum level corresponded to the next spectral line, and so on.

"Now comes the most surprising revolution of all, one so subtle that we have not yet even recognized it. The mathematical tools appropriate to the new view of reality had been developed by Newton and Leibniz in the eighteenth century, by Riemann and Lobachevsky in the nineteenth, and by many other mathematicians. My point is this: Look at the equation developed by the German physicist Erwin Schrödinger. It describes the energy states of the hydrogen atom in terms of the forces that bind its electron to its nucleus. Look at it!"

$$-\frac{\hbar^2}{2m}\nabla^2\Psi - \frac{e^2}{r}\Psi = E\Psi$$

I wasn't sure what Canzoni expected me to see. I stared at the equation, as though expecting it to sprout electrons and a nucleus. It stared back at me from the chalkboard, opaque, mysterious. As a mathematician confronted by such a formula, I was hardly in a better position than the so-called layperson. I may have known how to perform the operations indicated by the formula, but I had no idea what physical quantities the symbols represented. I may have known that the upside-down delta, the so-called *del operator* (∇), represents a fancy, multidimensional differential, but I did not know that the psi symbol (Ψ) represented the energy levels of the atom. That would be for a physicist to know.

Canzoni drew a breath at this point, as though she were about to dive into the canal below our balcony.

"You see, you could say that this equation *is* a hydrogen atom!"

Surely, I thought, there's more to a hydrogen atom than this. "What about the energy in the nucleus?"

"Well, true, there is more to the hydrogen atom than merely the interaction of electron and nucleus. There is the so-called standard model, a collection of equations that describes the energy interactions within the nucleus, of the quarks that make up the neutrons and protons. I am using Schrödinger's equation as a sort of symbol for the whole mathematical description. At the moment, Schrödinger's equation, together with the standard model, amounts to a hydrogen atom because there is simply nothing more to be said about it, as far as we know. Even if there is more to be said, I must ask, Do you understand what all of this means?"

I didn't, quite.

"These mathematical equations, including the ones we may not yet know, describe the energy relations within a hydrogen atom. The equations themselves are not energy, however. They are just equations. You could call them information systems in the sense that together they provide all the information you could ever want about a hydrogen atom. There is nothing else, in effect. You could say that not even the energy is real. Only the information about its behavior is real."

"Is this what you mean by the disappearing elephant?"

"Yes. The cosmos is an elephant. Examined closely, it disappears. It becomes its own description."

The morning was turning into afternoon, heat had invaded our balcony, and Canzoni sat down to fan herself. This seemed an ideal time to ask her about something that had always bothered me.

"I have often wondered about the ultimate structure of atoms. You have outlined what I call the *regress,* wherein matter is composed of atoms, atoms are composed of nuclei and electrons, nuclei are composed of neutrons and protons, and these, in turn, are composed of quarks and other things, I suppose. Does it all come to an end, or does this structure go on forever?"

I didn't want to derail Canzoni's train of thought but I couldn't resist asking. She responded, "The theoretical physicist Steven Weinberg thinks that what you call the regress must come to an end. He believes that soon we shall come to an end of formulas and know everything there is to know about physics. The cosmos will be founded on a finite set of axioms, themselves residing in the holos. But I do not share this view. It comes down to this question: Will we go on making substantially new discoveries about the cosmos forever, or will the process terminate someday, as Weinberg claims?

"This question has many profound implications for us. Why should there come an end to discovery? Why should the cosmos be Weinbergian, governed by a set of axioms, which, when compared to the holos itself, is almost insignificant? Why shouldn't the cosmos be governed by the entire holos? I have reason to believe that there will be no end to discovery, at least no end to the things that need discovering. In fact, what may be the biggest piece of all remains, for the present, undiscovered.

I was beginning to feel increasingly strange, as if our conversation were altering the reality around us. The balcony, the blackboard, and the coffee mug had taken on a temporary, ephemeral quality, as though about to disappear. Canzoni had started to tremble.

"Are you all right?" I asked her.

She fanned her face nervously with her hand.

"It's the *scleroso*. Heat is bad for it. You know where my office is. Would you get something? It is a bottle lying on the corner of my desk. Bring it here, if you would be so kind." I strode briskly to her office, found the bottle, and returned with it.

"I am very embarrassed that you should see me this way. I have much pain at times and I become dizzy very easily." Then, abruptly, she stood up. "Perhaps I should eat something. Shall we have lunch?"

We went to a small *ristorante* with outdoor tables not far from the Rialto. Canzoni appeared to be more cheerful now, talking animatedly about the Italian contributions to science and mathematics over the past five centuries. From Fibonacci to Fermi, Italy had contributed more than its share to European science. Gradually, however, her cheerfulness drained away as she told me about her life as a young physicist at CERN in Geneva.

There, she had watched with growing excitement as the giant accelerator revealed new phenomena in the spiraling tracks of subatomic particles. She had seen pair creation, particles and antiparticles congealing from the energy of the great accelerator beam. She had been a firsthand witness to the evanescence of the very substrate of reality. What had brought her career there to an end? She grew somber.

"There was a man there, a man whom I trusted. I will even confess, just between us, that I had certain feelings about him. At the same time, I was framing a kind of hypothesis about the *Mondo Mathematica,* as I called it then. But when I shared these ideas with my friend, he became most unfriendly, laughing in my face and calling me a fool, in so many words. At the time, I was shocked and hurt, but lately, with the years, I have come to realize that the ideas had perhaps frightened or threatened him in some way. After a few months, someone submitted a performance review on my behalf, claiming that I had not published very much during my time there, that I was not pulling my weight, as they say.

"Was your so-called friend behind this, do you think?" I asked.

"Ah, who can say? In what they call 'big science,' a kind of bureaucracy enters the picture. There is tension, there is a kind of jockeying for power among those with the greatest ambitions. I left gracefully, I think, finding by pure luck this position at the Universita' Ca' Foscari di Venezia in the city of my birth."

Just then, her graduate student Emilio happened by, seeing us and joining our table. "*Buon giorno, professori.* Dr. Canzoni, I have the paper you requested me to get for your visitor." He handed Canzoni a small paper, which she promptly handed to me.

"I have never tried to publish this, for I know that it will not be accepted, even laughed at. But you are the ideal person to give it to. If you write about the ideas you have encountered, be sure to include it, for it will be my only chance to publish, in a way, my ideas."

The article was fairly short, worded more like a manifesto than a scientific paper. Indeed, she had not been trying to write a paper, despairing of ever establishing her ideas with anything like physical (let alone mathematical) rigor. In the middle of it, I found a thesis, stated in three parts:

THE CANZONI THESIS

1. The conclusions of mathematics, both known and unknown, apply in full to every object, abstract or concrete, that satisfies an axiom system.

2a. Some things in the cosmos satisfy an axiom system. (weak form)

b. Everything in the cosmos satisfies axiom systems. (strong form)

3. The cosmos is the truth intersection of everything in mathematics. (the superstrong theory)

"What do you mean by the truth intersection?" I asked.

"It is this. Think for a moment about a particle. Its behavior is governed by certain mathematical laws. Its position, the time that it exists, its momentum and energy are all governed by one equation or another, expressing the operation of these laws. One could say that the laws that apply to that particle intersect or come together in that particle.

"Now why the particular laws we have so far discovered should apply to the cosmos and no others is a deep question. Whether or not other laws remain to be discovered, I would say that each particle and each wave must manifest where it manifests and must behave as it behaves because of the holos. You see, the holos contains all of mathematics, both what is known to us and what remains to be discovered—by far the greater part. In fact, you may be sure that there is no end to mathematics, whatever the situation might be for physics.

"I picture it this way. The fundamental constituents of the cosmos, whether they end with quarks or not, satisfy not just the axioms that Weinberg envisages, but also many others. How can we possibly have the nerve to say that we know all the laws or axioms satisfied by a hydrogen atom, for example, when we don't even know the tiniest fraction of what is in the holos? What I call the truth intersection of the holos is the set of all axioms that apply to a hydrogen atom and other things."

"I'm sorry," I said. "I don't quite see what actually causes a hydrogen atom to manifest itself."

She looked at me with that same sad expression on her face. "To be frank, neither do I. Something very important is missing from physics, I would say. But a final theory of everything, in my view, is more likely to resemble the sketch I have just given than the current picture of standard physics."

"What, then, is missing from physics?" I asked.

"It is not easy to say this because it is more outlandish than what I have already shown you. What is missing from physics?" She drew a deep breath. "Mind."

I had sudden misgivings, as if I had spent the past two days with a madwoman and hadn't even suspected it.

"Mind?"

"Physics deals with what we call 'matter.' Until the beginning of this century, there seemed to be no room in it for mind, for mental phenomena. But then came quantum mechanics, triggered in part by Balmer and his discovery of the formula for the wavelengths of the hydrogen atom. Niels Bohr and the Copenhagen School then pressed the new view of matter to its logical conclusion. For Einstein, it was not really the discreteness of matter or energy that bothered him, but something quite different, a factor that was quite new to physics: nondeterminism."

Her voice trailed off, as if she had forgotten what she was going to say next.

"What sort of nondeterminism?"

"Random behavior. There are various experimental setups that force a fundamental particle such as a photon to make a choice of which path to take from its source to a detector of some kind. The choice, as far as the Copenhagen School is concerned, is completely random."

"Do you mean that the photon understands it has a choice?"

"Not at all—at least, not if I were to speak for the larger community of physicists. It's simply unpredictable, in principle, which channel the photon will take. Einstein fought this conception until his dying day, yet quantum mechanics is one of the most successful theories physics has ever known, at least so far. And there is more, much more."

She got up to stretch, Emilio and I rising with her. "I think," she said, "we had better go back to the office. I will explain more about the matter of mind on the way."

It was difficult to hear her amid the babble of voices and cries of vendors out on the pavement beside the canal. I had to hold the tape recorder over her shoulder to catch her voice. It wasn't until I got back to my hotel room that night that I finally heard it all.

"Have you heard of the Nobel physicist Eugene Wigner? Yes? Well, about 40 years ago, he wrote a very interesting little essay that you should be aware of. It is called 'The Unreasonable Effectiveness of Mathematics in the Natural Sciences.' He said that there is simply no rational explanation for why mathematics plays such a crucial role in physics, no reason why it should be so useful, yet it is. You see, the great majority of physicists simply accept what Einstein or Bohr or someone else has told them, and they happily apply the theories in their laboratories or on their chalkboards. Also, many of them, most I would say, do not once stand back and shout, as Wigner did, so to speak, 'My God! What is this elephant doing here?' "

People stopped on the street to watch us pass. Emilio smiled at them ingratiatingly, but Canzoni seemed not to notice.

"As I am sure you know, there is something else going on in quantum mechanics. It seems to involve human conscious awareness as a key ingredient. It turns out that quantum mechanics is most successful when it assumes that there is no way to isolate the observer from the experiment. Unless certain phenomena are observed, they just don't happen."

"What sort of phenomena?"

"Suppose you send some photons from a source in the direction of a pair of slits that are very close together. If you do not interfere with the photons, they will interfere with each other, so to speak. On a screen behind the slit, you will see an interference pattern where the photons have acted like waves to either cancel or reinforce each other, depending on where they arrive at the screen. Yet if you arrange to observe which slit the photons go through, you destroy the interference pattern. One presumes that your awareness of the photons is what alters their behavior. That, at least, is what some physicists believe.

"For example, Wigner believed it—and not only did he believe it, but he found in such quantum phenomena a possible source of entirely new material for physics: consciousness."

"Do you mean some physicists have actually worked out a theory of consciousness?"

"Unfortunately not. The project may be hopeless, but here is what I think. There can be little doubt that consciousness involves a completely different order of physical reality than ordinary matter and energy. You may be sure that consciousness will never develop in a computer, no matter how it is programmed, because computers are designed to screen out the very phenomena on which consciousness may depend. Computers are built to resist errors, or random noise, quantum fluctuations in the states of billions of tiny transistors.

"Computers aside, we humans have consciousness. Do our brains involve just neurons and pulses traveling between them? No, there is something else going on that is completely beyond our comprehension at the moment.

"What's missing from physics is precisely what's missing from our present picture of the brain. What's missing from physics is the mechanism or force or . . . whatever you might call it, that brings mathematics to bear on things, so to speak. What's missing is the thing that brings phenomena into and out of manifestation, like thoughts that come and go. For they are like thoughts that come and go, you see . . ." Her voice trailed off. "I'm sorry to be so vague," she said. "Pygonopolis might call it the *menos,* which is Greek for 'will' or 'spirit.' "

She had begun to tremble again. Luckily, we were back at her building. We climbed the dark stairway in silence. Not until I sat across from her desk did she finish the thought. "It goes like this, very roughly. There is something like consciousness that permeates the cosmos. To prevent premature conclusions or unwelcome analogies, we will give it a neutral name, say the *menos*. The nature of its existence is completely different from matter and energy, yet it works like a *materium primum,* the basis for many phenomena. There is, accordingly, a *menos* that pervades everything, and there is in each of us a little piece of it, our own consciousness, the one to which our brain plays host. I am not saying that the *menos* is unified in any sense, only that the phenomenon pervades reality as we have yet to know it."

"Does this overall consciousness have something to do with mathematics?"

"The great consciousness is the missing ingredient, I think. Only through or by this consciousness is mathematics deployed, somehow, in the cosmos. We may even have to reconcile ourselves to never knowing."

"Why?" I inquired.

"Because of what I call the 'quantum curtain.' It hangs between us and the deeper phenomena. The curtain consists of the essentially random behavior of fundamental particles. Their behavior is unpredictable, in principle. Yet there could be something behind the curtain, something that determines quantum events yet is not to us predictable. Behind the quantum curtain lurks the overall consciousness, perhaps. If the holos is anywhere at all, it is there."

My plane was to leave shortly. I did not understand all her

remarks, at least not very clearly, yet I would clearly have something to ponder on my flight to England. Given her thesis, Canzoni could only be described as a hard-core Pythagorean. But the thesis, fascinating as it was, had been somewhat overshadowed by a rush of imagery, always the sign of a groping mind. There was the invisible elephant, the one into which the cosmos disappeared when examined closely, there was the *menos*, or vast field of consciousness that supposedly pervaded everything, and there was the quantum curtain behind which we could not peek. My quest had taken a sudden and unpredictable turn. What would my next host, Sir John Brainard, have to say about it all?

PART IV

The Engines of Thought

CHAPTER 7

Horping Zooks

Oxford, England, June 29, 1995

As my train sped north to Oxford from London, the phrase "England's Green and Pleasant land" kept ringing through my head. Was it from William Blake? It was certainly a green and pleasant day. The landscape would occasionally roll, lined with stone fences, dotted with sheep. Elsewhere, the train descended to the valley of the river Thames, where pleasure craft and occasional barges dragged wakes through the water.

When my train pulled into Oxford station, I descended to the platform knowing full well my host would not be there to greet me. He was, after all, Sir John Brainard, Senior Fellow of Merton College and renowned for his comprehensive grasp of mathematics. I had been told not to expect Brainard to meet me, as (a) he was now extremely old, and (b) nobody should expect to be met by such a luminary, in any case. I therefore hired a cab to carry me into the city, where I finally met the great man by the doors of his college. As he stood talking to the porter, I hovered nearby, suitcase in hand, hoping to be noticed. Finally Brainard turned. "Good heavens! You're not the Dewdney chap I've been expecting, are you?"

I tried desperately to think of something clever to say but succeeded only in mumbling.

"I gather you are he, then." Sir John introduced me to the porter, explaining that I could leave my bag with him while we took a stroll by the Cherwell (a river that flows into the Thames at Oxford).

"I've taken the liberty of booking you into the Churchill Hotel, if that meets with your approval. My younger colleagues all seem to feel that the way to treat a guest after an arduous journey is to rush him off to the nearest public house and ply him with pints until he can barely stand. Hardly the thing for a clear head. Besides, you look like the sort of chap who prefers fresh air to the smell of smoke and hops."

Brainard led me around the college and out onto a parklike expanse of grass bordering the Cherwell. A few people wandered along a path there, turning occasionally to watch a punter make his way slowly upstream. I stole a glance at Brainard. He was certainly quite old, but of that indeterminate age that belongs to legends. His hair was a huge shock of unruly white, and his eyebrows had grown to such lengths that I felt an irrational urge to take scissors to them. Here he was, beside me, said to be the last person with a fairly complete understanding of all mathematics and author of *The Mathematikon.* It was said that, besides his prodigious output of articles on every conceivable area of mathematics, his shelves were lined with unpublished papers, any two of which would make the reputation of a postdoc starting out.

"I must say, I am in full sympathy with your quest," he said in a kindly way. "But you have posed something of a false antimony, in my view, by asking whether mathematics is created or discovered."

"I'm sorry?" We had barely gone beyond pleasantries, and I was already feeling a bit lost.

"To say that mathematics is discovered presupposes that it already exists in some sense. Yet to say that it's created implies that it did not previously exist. Yet how one could possibly determine the question of existence, let alone preexistence, is frankly quite beyond my powers—and yours too, I suspect."

He turned to stare at me. His eyes were a shockingly pale blue, and somewhat watery. He meant what he said.

I tried another tack. "But couldn't one say that mathematics seems to have an independent existence?"

"Ah, you say 'seems.' That's progress. But how to answer such a question? What sort of existence are you talking about? Physical existence?"

"Well, I suppose not," I said, "but the intimate relationship between mathematics and physics, the very power that it lends to physics, makes me wonder whether mathematics exists, in some sense, behind the scenes, even as physics exists before our eyes."

"Ah, the spirit of Oxford has already begun to permeate you," Brainard retorted. "In time, should you remain among these dreaming spires long enough, you would cease to have anything remotely resembling thoughts."

Brainard gave a loud laugh that startled a child and her nanny. It had taken me this long to realize that he was playing with me, testing me. Yet I sensed a certain nervousness on his part, as though he might actually be a rather shy person. I put the matter as plainly as I could: "You cannot have lived so long or produced so many significant results without reflecting deeply on the question, or at least on how the question might be posed."

"Well done! Passion and patience, a touch of flattery, an appeal that touches me. And so we can start. Is there something like a place where mathematics exists? Well, it certainly exists in our minds. I might add, in case you think that a poor sort of existence, that things of the mind are as real as any other part of the physical world. Unless you are prepared to invoke a divine or mystical element, anything that has physical effects must itself have a physical existence. And the mind, as we all know, can have physical effects of the most profound kind. Thus, in an indirect way, I can say that wherever else mathematics may be, it is also part of the physical world."

"But could you not say the same thing of unicorns?" I asked.

"And of lions, as well," replied Brainard instantly. "But I am referring to the reality of concepts and mental operations, not physical things such as lions and unicorns. And I should warn you that not all concepts and mental operations are equal. However mathematics may be said to exist, I will insist on the mind as its principal theater."

I found this reply somewhat disappointing. To move him back to the idea of an existence that was independent of the mind, I told him about Pygonopolis and his concept of the holos.

"The holos!" exclaimed Brainard. "Yes, well, it sounds terribly Greek. In spite of the fact that we are surrounded by a university that has specialized in Greek for many centuries, I haven't the faintest idea of what the word means. Is it from Plato?" he asked.

I replied that the term meant "the whole" and was recently coined to name the world that Pythagoras thought underlies all of existence. I pointed out that Pythagoras thought the world was actually made out of number.

"Yes, well, we've all heard that chestnut," retorted Brainard somewhat impatiently. "But since I have no idea what the word *holos* means, I shall not use it. I have the feeling that you want me to invent a word like Mathsland, in the end guaranteeing nothing more serious than Wonderland, as in *Alice in* . . .

"By the way, I hope you realize that Oxford was once home to Lewis Carroll."

I did, vaguely.

"His real name was Charles Dodgson, a fellow over at Christchurch, just behind us. Not a bad mathematician, as mathematicians go, but no serious work to speak of. He was certainly fond of games and, for that matter, of little girls. The theory is that he sublimated his urges and produced, in consequence, a marvelous fantasy world where his heart could dwell on or in Alice, forever. The original Alice was a daughter of Dean Liddell of Dodgson's college. He treated her with the utmost respect, although he was known to have photographed her in the nude."

The conversation seemed to be taking an unhealthy twist. I made so bold as to interrupt Brainard.

"If not Mathsland, then what?"

"In my casual reading, I have encountered only one attempt to describe mathematics as a separate place. Have you heard of World Three?"

"I don't even know what Worlds One and Two are!"

"Well spoken. World Three is the invention, if not the discovery, of two fellow knights of the realm: Sir John Eccles and Sir Karl Popper. Together they collaborated on the book called *The Self and Its Brain*. In it, these eminent thinkers, one a neurophysiologist, the other a philosopher, outline three worlds in an attempt to describe the very special role played by human consciousness in the physical world."

This came as a shock. I chided myself for not knowing this already. I could even remember seeing reviews of the book, but because I was not particularly interested in neurophysiology, I had skipped them. At the same time, mention of the word "consciousness" instantly brought Maria Canzoni to mind, along with her claim that consciousness played a special role in physics. Somewhat humbly, I asked Brainard about the three worlds of Eccles and Popper.

"World One consists of physical objects, the sorts of things you can see, feel, or push about. You could call it the world of physical reality. World Two consists of states of the human mind, both conscious and unconscious. This world can have a direct impact on World One, rather in the vein that we have just been discussing. States of mind, particularly willed acts, can have a direct effect on the physical world."

He paused for a lengthy period, causing me to push him a little. "What then is World Three?"

"World Three, if I have this right, consists of all products of the human mind, from music to mathematics. Ah, I see by the way your face lights up that I am onto something here! I suppose it is Popper the philosopher who struggles with the form of existence of objects in World Three. Take music, for example. Is music the notes on paper, the sounds made by a symphony, or the groove on a gramophone record? It is all of these things, of course, but none of them, as well. World Three is real because, whatever form music ultimately takes in the real world, it also has direct physical effects. In a sense, it causes the several tons of musicians and instruments in a large orchestra to bellow forth with a blast of Beethoven."

Amazingly, it hadn't occurred to me until now that mathematics wasn't the only field with objects that had no ultimate definition. I remembered Pygonopolis's attempts to describe the ultimate reality of number, al-Flayli's efforts to describe the ideal circle.

Brainard continued, "Now you may call the place where mathematics exists World Three if you like, but I find the concept wanting in several respects. First, it puts mathematics on an equal footing with the static patterns that form its subject matter. Whatever else music is, it must be a static pattern. World Three only makes sense as a philosophical concept if you include all possible strings of binary bits. As everyone now understands, thanks to the computer age, these are capable of encoding all music, all art, all literature, along with all sorts of nonsense patterns. Who can say what is art, whether already realized or potential—especially these days," said Brainard with a dry chuckle. "Something out of Borges, I suppose."

I let the last remark slide. "Do Popper and Eccles claim that World Three has an independent existence?"

"Indeed, they do. My point is only that they do not recognize that mathematics exists at a higher level, in a sense. It is *about* the strings themselves, among other things.

"Frankly, I'm much more interested in how we do mathematics. Any examination of mathematics and the kind of independent reality that it may or may not have must begin with the mind. What you may not realize, however, is that by *mind,* I mean a great deal more than just the human mind."

My ears perked up.

"Mental operations of the sort that make mathematics possible are not confined to human beings, as you will see. Indeed, a major leitmotif of mathematics in this century has been the detachment of mathematics from the human mind, if I may put it so crudely. This development has arisen in part from the axiomatic method whereby we have put much of mathematics on a more or less unassailable footing. I will touch on other contributing factors later, but let us concentrate for the moment on axiomatics. Our story will begin with the central phenomenon, mathematical thinking."

"We sat on a bench with the late afternoon sunlight streaming through the trees and illuminating a heavy pipe, which Brainard produced from his vest pocket.

"Rare thing, sunlight!" he remarked, as he lit up an aromatic mixture. "Doesn't bother you, I trust?"

I shook my head.

"Mathematical thinking is not at all like ordinary thinking," Brainard continued. "Your thoughts are focused in the most extraordinary way on objects so extremely simple, so bereft of detail, that you understand them fully. There is simply nothing more to understand about a number than the quantity it represents. This thought of quantity, this concept, has a special quality, that it is identical in every person who understands it. To be sure, each individual mind plays host to a plethora of idiosyncratic associated ideas that flit about the central fire, but they play no role in the action that such a thought has.

"For example, some mathematical words are quite ordinary words, such as *group*, *normal*, *function*, and so on. But in their mathematical use, they have little or nothing in common with their ordinary use, so people encountering these words for the first time in a mathematical context are apt to import meanings from ordinary life, thus clouding their understanding."

An amusing thought struck me. "Are you saying," I asked, "that most people have trouble understanding mathematics because it's too simple?"

"Capital idea!" Brainard slapped his knee, sending pipe ashes tumbling onto his trousers, an accident he didn't seem to notice.

"Let me give an example of the incredible simplicity that lies in the heart of mathematics. I'm going to set up a system of axioms in which all the key concepts are words you have never heard before, thereby avoiding any possibility of confusion. By the way, if some of the words I use sound as if they were invented by Lewis Carroll, at least I come by the art honestly.

"The subject is blorgs. What is a blorg? First, a blorg consists of zooks. Second, you may horp one zook with another and the result is always a zook."

My mind reeled, as though I were an undergraduate once again. Here were three completely unfamiliar terms, and Brainard had barely started.

"I will allow that a blorg is something that consists of zooks, but I have some questions about horping. What exactly is horping? Can you give me an example?"

"I have already told you: Horping is the process you apply to two zooks to get a third. Examples are extraordinarily easy to produce. Mind you, I haven't quite finished defining a blorg, but if you will allow me to call the thing defined so far a semiblorg, here is an example."

Brainard, who had kept a pen and pad at the ready all this time, jotted down a little table for my inspection.

	a	b	c
a	b	c	a
b	c	b	a
c	c	a	b

"You could call this the horping table for a particular semiblorg. It's all there. In this particular case, the zooks are called a, b, and c. And if you want to horp zook a with zook b, for example, you go to where the a row meets the b column. At the intersection, you find c. In other words, if you horp zook a with zook b, you get zook c. What could be simpler?"

To be sure I understood Brainard, I asked him whether every semiblorg could be expressed by such a table.

"Oh, yes indeed," he replied. "At least the finite blorgs. You will notice that I have said nothing about the number of zooks that may be in a semiblorg. The axioms I am constructing allow for either a finite number of zooks or an infinite number of them. I should also mention that we are not confined to tables in the matter of producing actual semiblorgs. For example, I might introduce a special notation for horping, say, a pound sign. So the following little equation is equivalent to looking up the result of horping zook a with zook b in the table:"

$$a \# b = c$$

I did a quick calculation in my head. "In other words, if you listed eight more such equations, you would have specified this particular semiblorg as completely as the table does?"

"That's correct. There are just nine horpings possible in the table, after all. To get another semiblorg, simply make a new table. Continue using alphabetical letters, as many as you like, and fill the tables with those letters in any way that strikes your fancy. The result will always be a semiblorg. Unfortunately, the same thing will not be true of blorgs, so we had better press on in order to complete the blorg axioms.

"Ah, by the way: The introduction of the special notation for horping, the pound sign, illustrates the power of good notation. In the coming development, if all we had was the table representation for blorgs, we should be practically helpless when it comes to probing mathematical structure or to expressing ideas we might have about blorgs. You will see how handy the pound sign is. There is nothing like effective notation for guiding the engine of thought along the rails of discovery.

"You will recall that a semiblorg is what we have defined so far, with zooks and horping. A blorg, then, is just a semiblorg with a few more axioms. The next axiom concerns a very special zook, the one I call a gadzook."

Brainard maintained a poker face, so I suppressed my smirk.

"The gadzook has the property that when you horp it with any other zook, you get the same zook back again. Here. Suppose zook z happens to be the gadzook. Then for any other zook, say a, we have this:

$$z \# a = a \quad \text{and} \quad a \# z = a$$

"Now besides the gadzook, a blorg has something else going

on in it, something that justifies the introduction of the gadzook. For every zook in a blorg, there is an antizook. Moreover, when you horp a zook with an antizook, you always and invariably get the gadzook."

"Sounds like physics," I remarked.

"Perhaps," said Brainard with an undertone of irritation, "but a mere coincidence. By a different choice of words, I could make it sound like knitting, but it would be exactly the same thing."

"Can a blorg have more than one gadzook?" I asked.

"Excellent question," enthused Brainard. "Let's take the matter up right now. Even without the next and final axiom, we can probe your question. Suppose that a blorg, as so far defined or axiomatized, can have two gadzooks. Call them z and z'. By the gadzook axiom, we know that when you horp the first gadzook with any other zook, you get the other zook again. Thus, when you horp the gadzook z with the gadzook z', which is after all still a zook, you get z' again. Here it is:

$$z \# z' = z'$$

"By the same axiom, we can reverse the order of horping and get the same result.

$$z' \# z = z'$$

"But when we apply the same axiom to the other gadzook, z', and horp it with the gadzook z, we must have

$$z' \# z = z$$

"This means that z and z' must be the same zook, for they both equal the same thing."

For some time, I had been aware that other axioms would be lurking in the background, not just the axioms for a blorg that Brainard was laying out but a whole suite of axioms that concerned deduction, the apparatus that Brainard was using without being explicit. "Aren't you invoking an axiom outside the system you are defining? Aren't you using an axiom of equality? You know, it goes back to standard Euclid: 'Things equal to the same thing are equal to each other.' "

"Hmmm," came the response. "I was hoping you wouldn't bring that up. Yes, indeed, I have just used the so-called axiom of equality. And yes, it belongs to another set of axioms that is more or less universally applied to all mathematical reasoning. But I am coming to that later, as well. Please do let us stick to blorgs. I have answered your question about the uniqueness of the gadzook and am about to deliver the final axiom. One never knows when Olympus will pour the deadly draft for this old Socrates.

"Finally," he continued, "in a blorg, you may horp three or more zooks in a row without worrying about the result. Suppose I wrote down the operation of horping three zooks in a row:

$$a \# b \# c$$

"What does this mean? Because you can only horp two zooks at a time, you will have to indicate, perhaps by a bracket notation, which pair of zooks you want to horp first. We thus have two choices about how to carry on:

$$a \# (b \# c) \quad \text{or} \quad (a \# b) \# c$$

"The final axiom is simply this: In a blorg, it doesn't matter which way you go about horping three zooks, the result is always the same."

$$a \# (b \# c) = (a \# b) \# c$$

"That looks a little peculiar," I said. I was beginning to enjoy the game of egging him on. "Why on earth should we worry about the order of horping?"

"All I can say for the moment is that it's an essential feature of blorgs. In a moment, you will see how very useful this last axiom is. Because this system of axioms has been detached from any and all applications, I cannot say more about the significance of this idea.

"In any event, the axioms I have given you—the zooks, the horping, the gadzook, the antizook, and the law of triple horping—all define what a blorg is. The axioms are complete, and I am now ready to explore what theory there may be waiting for us. In this endeavor, I must remind you that we began this discussion with my remark that mathematics is difficult because it is so simple. Everything you need to know about blorgs is explicit or implicit in the axioms I have given you. As you enter the Spartan spirit of this world, a certain purity of thought becomes evident."

"Before you continue, Sir John," I interrupted, "could you give just one example of a working blorg, so to speak?"

"How rude of me not to," exclaimed Brainard. He scribbled a new table in his notepad and showed it to me:

	a	b	c	d
a	b	c	d	a
b	c	d	a	b
c	d	a	b	c
d	a	b	c	d

"Now this blorg has one more zook than the semiblorg I showed you earlier; otherwise, it appears rather similar. But if you examine it closely, you will see that it is more highly structured. The gadzook in this case is the zook called d. Do you see how it simply reflects whatever zook it is horped with? Also, each zook has an antizook. The antizook of b, for example, is obviously c because

a # c = d,

where d is the gadzook. Now there are two things I wish to say about this particular blorg. First, I will open the gates to the real world by showing you where this particular blorg might be found. Then I will show you an interesting structure inside this blorg. That structure, in turn, will pave the way to the bit of theory I wish to develop.

"If you take a square and rotate it by 90 degrees, you will get the same square over again. Let us agree to call that rotation *a*. Now when I draw the rotated square, it looks just like the original one, so we will label one corner to show what has happened." Brainard drew two squares on his notepad:

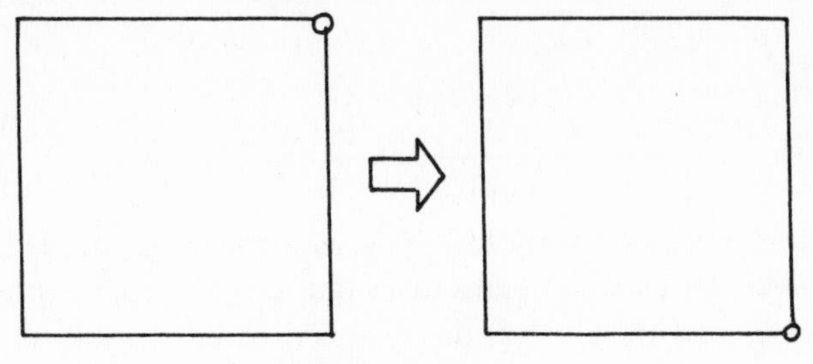

A Rotation Is a Zook

"Now a rotation of 180 degrees also leaves the square unchanged, as does a rotation of 270 degrees. We will call these rotations *b* and *c*, respectively. These are the zooks in that table, if you like, and the final one, *d*, is the null rotation. You do nothing to the square. We could therefore call our example the rotation blorg for a square.

"Do you see how horping works? Simply follow one zook by another. If I follow the 90-degree rotation *a* by a 180-degree rotation *b*, I get the zook a # b = c. Also, each zook has its antizook: For example, the rotation *a*, followed by the rotation *c*, gives you the null rotation *d*, the gadzook of this particular blorg."

There was no doubting that the rotations of a square amounted to a blorg, as Brainard said. Nonetheless, I couldn't help wondering about the relationship between the abstract example, a mere table, and the concrete one, the rotations of a square. "Doesn't this mean that blorgs can be considered real things, at least insofar as this particular one reflects certain realities of the physical world?" I asked.

"Certainly. If you take an actual square made of cardboard, for example, and proceed to rotate it by the amounts mentioned, you will automatically express this particular blorg and fall under the axioms. More than that, you will also be constrained by all the implications of the axioms of a blorg, including the theorem I am about to prove for blorgs. Just look at the zooks *b* and *d*, for example. Together they constitute a blorg!"

It was true. Here is the little table that Brainard drew:

	b	d
b	d	b
d	b	d

"Because this subset of the zooks of the blorg itself forms a blorg, we call it a subblorg. Notice that this subblorg has just two elements, while the blorg itself has four. As you know, 4 is a multiple of 2, and that brings me to the theorem I wish to prove." Brainard wrote the theorem on his notepad:

> Theorem: If B is a blorg and C is a subblorg of B, then the number of zooks in B is a multiple of the number of zooks in C.

"To prove this theorem, we will show that it is possible in every case to divide the blorg B into equal-sized bunches called coblorgs. The coblorgs will all have the same number of zooks in them as B, and they will also be nonoverlapping. It will then follow that the number of zooks in B must equal the number of zooks in C, multiplied by the number of distinct coblorgs. That, in turn, will mean that the number of zooks in B is a multiple of the number of zooks in C."

Brainard drew a little diagram to illustrate the proof procedure. This was a purely schematic figure, because blorgs are algebraic objects, not geometric ones. The large rectangle represented a blorg, and the smaller rectangles inside it represented the coblorgs.

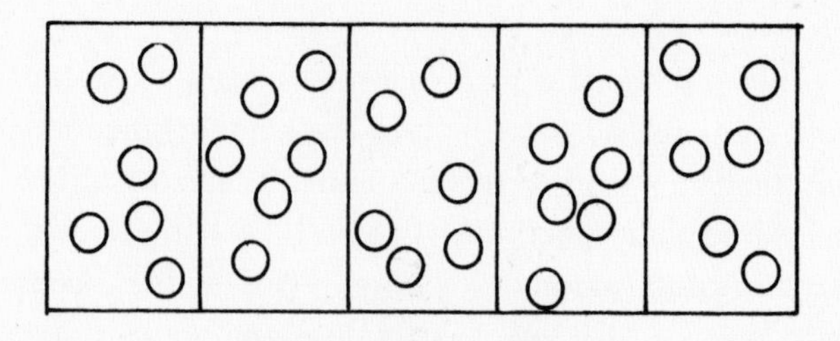

A Blorg Divided into Coblorgs

"Now a coblorg of C is simply the set of all zooks you get by horping a specific zook in B by every zook in C. So if I take a particular zook b and horp it on the right with every zook c from C, I will obtain a whole set of zooks, which I will write as follows:

b # C

"The coblorg b # C consists of all zooks of the form b # c, one for every zook c in C. How many zooks do you suppose there are in b # C?"

"I don't know," I replied. "I would guess the same number of zooks as in C. It would all depend on whether it was possible to horp b with two zooks c_1 and c_2 in C and end up with the same zook. If so, there could be fewer zooks in b # C than in B."

"Well said," chortled Brainard, obviously enjoying the game. "It's time for a lemma."

A *lemma* is just a small theorem that paves the way for a large one by proving a result that the larger theorem will need. In this case, Brainard would have to prove the statement he wrote next on his notepad:

Lemma (the law of cancellation): In any blorg,
if b # c = b # d, then c = d.

"You will recognize that you have really asked whether a law of cancellation prevails for blorgs. In the matter at hand, we have your question whether b # c_1 = b # c_2 implies that $c_1 = c_2$. If our lemma is true, it obviously does."

"I presume the lemma is easily proved," I murmured, hoping he wouldn't be too long about it.

"Quite, quite. We simply begin with the given statement b # c = b # d and apply the antizook axiom, the one that permits us to horp any zook by its antizook. By the way, we have as yet no notation for antizooks, so let us denote by b' the antizook of b.

b' # (b # c) = b' # (b # d)

"Now you may recall that I promised that triple horping would come up, and here it is. Thanks to the last axiom, however, we may rewrite both sides of this equation as follows:

$$(b' \# b) \# c = (b' \# b) \# d$$

"Of course, when you horp a zook with its antizook, you get the gadzook, so we can rewrite the equation again . . .

$$z \# c = z \# d$$

and, because the gadzook leaves every zook unchanged under horping, we have the final result, thus proving the lemma."

$$c = d$$

"That's a lot of work for such a simple result," I ventured.

Brainard looked at me strangely. "I think not. Please don't forget that we shall never have to prove it again. We will add this little lemma to our knowledge base about blorgs. Obviously it's a very useful little result and properly speaking should not be a lemma at all, but a minitheorem that, in spite of its size, should come early in any theoretical development of blorgs."

"Well, I withdraw my remark, in that case," I said. "If I may recapitulate, you have shown, via this little theorem, that when you form the coblorg b # C, the number of zooks in the coblorg is exactly the same as the number of zooks in the subblorg C."

"Yes," he replied with an air of satisfaction. "And now comes what you Americans call the 'crunch.' "

"I'm not American," I said.

"I beg your pardon," said Brainard. "Old age, you know."

"The crunch comes when you ask what relationship holds between two of those sets, say b_1 # C and b_2 # C. In particular, what if these two coblorgs have a zook, call it d, in common? In such a case, because d is in b_1 # C, it must have the form b_1 # c_1, for some zook c_1 in C. However, because d, the zook that lies in both coblorgs, is also in b_2 # C, it must also have the form b_2 # c_2 for some zook c_2 in C. We can therefore write

$$d = b_1 \# c_1 \text{ as well as } d = b_2 \# c_2$$

"By the equality rule we discussed earlier,

$$b_1 \# c_1 = b_2 \# c_2$$

"This time we simply multiply both sides on the right by c'_1, the antizook of c_1:

$$(b_1 \# c_1) \# c'_1 = (b_2 \# c_2) \# c'_1$$

"Good gracious! It appears that the triple horping axiom will come into play again:

$$b_1 \# (c_1 \# c'_1) = b_2 \# (c_2 \# c'_1)$$

"And now,

$$b_1 \# z = b_2 \# (c_2 \# c'_1)$$

"At this point, we have b_1 # z on the left, which is simply b_1, because z is the gadzook. But on the right, you will notice that c_2 # c'_1 is a zook in C, for C is a subblorg and therefore a blorg in its own right; when you horp any two zooks in C, you always get a zook in C. So now we have

$$\begin{aligned} b_1 &= b_2 \# (c_2 \# c'_1) \\ &= b_2 \# c_3 \end{aligned}$$

Here, to simplify the notation, and without doing any violence to the argument, I have replaced c_2 # c'_1 by c_3. Evidently b_1 must belong to the coblorg b_2 # C, because it may be expressed as b_2 horped with a zook in C, namely c_3.

"The logic rolls inexorably along. It now follows that if you horp b_1 with any zook c in C, you get the zook b_1 # c, which must belong to b_2 # C, which the following lines of algebra show:

$$\begin{aligned} b_1 \# c &= (b_2 \# c_3) \# c \\ &= b_2 \# (c_3 \# c) \end{aligned}$$

"What does this last expression mean? Every zook in b_1 # C is also a zook in b_2 # C, because c_3 # c is a zook in C. So the coblorg b_1 # C is contained in b_2 # C. We may repeat this argument in the other direction to show that the coblorg b_2 # C is also contained in b_1 # C. This can only mean that the two coblorgs are identical, even though they are generated by different zooks: b_1 and b_2.

"So finally we have a proof that any two coblorgs are either exactly the same or entirely disjoint, with no element in common."

I could see the way to the end now. Brainard's argument formed all possible coblorgs of C, one for each zook in the blorg B. Any two coblorgs that overlapped even in one zook were entirely identical. Otherwise, two coblorgs were entirely disjoint. One could therefore divide the entire blorg B up into disjoint coblorgs, all of the same size, that size being the number of zooks in C.

Brainard appeared to have finished. "I wonder," he mused, "whether you know what's been going on with these blorgs and zooks."

"I've been feeling a little déjà vu," I replied. "It's the feeling of having been through this before."

"Well, you have. The subject is not really blorgs, but groups. We have not only provided the main axioms of group theory, but also proved one of the fundamental theorems of group theory. It is Lagrange's theorem, which turns out to be a major tool for probing all kinds of groups."

With this remark, it all came back to me. In my undergraduate courses, we had used other words for zooks and horping, but the end result was exactly the same. One of the main concepts of modern algebra, *groups* are a generalization of many number systems. For example, if one takes ordinary integers as zooks and ordinary addition as horping, the result is a group. In this case, the gadzook is 0, for 0 plus any integer yields that integer again. The antizook of an integer is simply its negative. For instance, the antizook of the zook 5 is –5, for 5 + –5 = 0.

Moreover, if you take all the rational numbers—namely, the ratios of integers, such as 3/7—and regard multiplication as the horping operation, you also get a group. Here, the gadzook is 1, and the inverse of a rational number such as 3/7 is 7/3, because 3/7 × 7/3 = 1.

The groups go on and on. If you consider all permutations of a sequence such as abcde to be zooks, you horp two permutations by applying first one then the other. For example, if one permutation interchanges the first two letters, and another shifts every letter one position to the right (bringing the last one around to the front), then the first permutation changes the sequence into bacde, and the second permutation changes bacde into ebacd. The gadzook is the null permutation, where you do nothing. Again, for every permutation, there is an antipermutation—namely, doing just the opposite.

"I have the feeling," I said, "that you did not convert the theory of groups into the theory of blorgs for idle reasons."

"My main point was to demonstrate the utter simplicity of mathematics," said Brainard. "In particular, recall the very beginning of our thought sequence. I gave five axioms for group theory. For what other subject can you lay down the entire foundation with utter precision in about 10 minutes of conversation? The answer is none."

Brainard had perhaps underestimated the time it had taken to lay down the axioms of group theory, but I took his point.

"The confusion you felt," he continued, "was due to your searching for other meanings outside of the axiom system being spelled out. There is no other meaning. What is a zook, after all? It is the thing of which blorgs are composed. It has certain properties, according to the axioms, but no others. Every blorg contains a gadzook, and every zook in a blorg has an antizook there. In addition, there is a very simple axiom that tells you it is all right to horp three zooks at a time. That's all there is to a blorg—excuse me, a group.

"Now I will certainly admit that once I had laid down the axioms, things got more complicated, but they remained, I hope, utterly clear. In fact, a further 10 minutes' conversation sufficed to prove an important theorem in group theory. We moved systematically, building up new theorems from old. For example, we proved the cancellation law on our way to Lagrange's theorem. Every step was built on the previous ones, following as a necessary conclusion and without guesswork of any kind. And so mathematics proceeds.

"Mathematical foundations are so simple as to be painful or boring to the ordinary person. What most people miss about mathematics is that no real development is possible without such simplicity. It is the simplicity of a musical scale, if you like, turned by degrees into symphonies of thought."

"Aha," I said, "then mathematics is created."

"It was only an analogy," said Brainard. "Wherever you may think the symphony comes from, it has qualities of harmony and melody. The harmony concerns the way mathematical ideas all dovetail, never contradicting, always cooperating. The melody describes the flow of ideas within a particular development, such as the proof of a theorem.

"Not so very long ago, there lived a notable mathematician of this university. His name was G. H. Hardy. Hardy believed very much in mathematics as an art akin to music or sculpture. It was the purest form of thought—so pure that Hardy would have nothing to do with any mathematics that might be applied to anything in the real world. Mathematics was the queen not only of the sciences but also of the arts, and he would have her be servant to none. At the same time, strangely enough, Hardy regarded mathematics as having an independent existence—out there, as it were."

"Out where?" I asked.

"I haven't the faintest idea," Brainard retorted.

I examined his expression closely, but Brainard stared into the western sky with a singularly innocent expression on his face. Then he turned to me, smiling impishly.

"I would say out there—in here." He tapped his head. "But before we get to that, let us be perfectly explicit about what the blorg example tells us. First, mathematics, at least in its foundations, is the simplest subject known to man. Everyone finds it difficult for this very reason, I think. People are always shocked when they discover this fact for themselves. 'Oh, my goodness,' they say. 'I had no idea!'

"The next thing our example illustrates is the generality of mathematics. As you know, there are a great many different mathematical objects that have the structure of a group, not to mention a number of physical systems in the real world. Every single theorem in the theory of groups applies to every single one of those objects, without exception. In what other field can you make pronouncements that affect untold numbers of structures, both known and unknown?

"Finally, I wanted you to see the axiomatic method at work. Recall how frequently I wrote down a little string of symbols. In this, we are skirting the edge of the greatest development, in my humble view, of twentieth-century mathematics."

Hearing this, I almost had the feeling of being present at a historical occasion. "And what development was that?" I asked.

"My dear boy! What development could it be but the mechanization of mathematics. You see, we have discovered, without necessarily taking full advantage of the discovery, that our very thought processes, at least as expressed in proofs, can be reproduced in the machines we call computers. Now I propose to take this topic up more fully tomorrow—if I live through the night, that is. Oh dear! It's getting late."

The sun was getting low in the sky, and I glanced unobtrusively at my watch as Brainard gazed through the trees at distant clouds. It was nearly nine o'clock! I had forgotten that England's high latitude prolonged the summer day.

"Why didn't you stop me from babbling on like that? We shall repair, with your permission, to my favorite public house hereabouts, the Trout."

We strolled out of the college grounds and presently found ourselves on High Street, heading for a stand of taxis near the town center. As we went, Brainard explained that his doctor had ordered that he eat as little red meat as possible. "As a result," he said, "I'm using your visit as an occasion to indulge in a monthly treat of steak and kidney pie, something they do awfully well at the Trout."

Brainard apparently had no car, living close to Merton College as he did. As we rode, he explained his favorite real-world group, the orthogonal group of rotations. The zooks were all possible rotations of a sphere about one axis or another, passing through the center.

"You can visualize this group," he said, "by taking a large ball, choosing an axis of rotation, and rotating the ball through a specific angle. Do this twice, and you have horped two zooks to produce a third zook." I found this somewhat astonishing.

"You mean two such rotations, no matter what the axes or the angles, are always equivalent to a third rotation?"

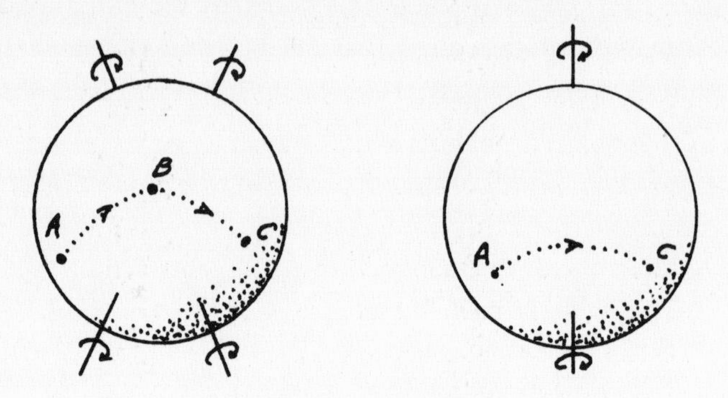

Horping Two Rotations

"Indeed," replied Brainard, somewhat smugly. "You can do a thousand such rotations in a row, it doesn't matter. The end result can be reproduced by a single rotation. Think about that. It isn't exactly obvious, yet it springs at us straight out of theory."

"What would you say," I asked him, "if someone rotated a sphere a number of times in this fashion and ended with an orientation of the sphere that could not be reached by a single rotation?"

"Such a finding would have to be translatable into group theory and would constitute a counterexample to a major theorem. It would be impossible, to my way of thinking. Impossibility in the mind implies impossibility in the real world in a case such as this."

"So you will concede," I asked, "that the world of mathematics, as you call it, is not without a certain influence on the so-called real world?"

"My dear chap, do let's be straight about this. I don't know what you mean by influence, but I think it is fair to say that certain logical impossibilities also amount to physical impossibilities. Let's just agree that the human mind has a certain ability to model reality, an ability honed by millions of years of evolution. We shouldn't have survived without a model that was essentially correct. In that model, we confront still ponds with perfectly flat surfaces that reflect our images, we see spherical bubbles, straight lines at certain horizons, and points everywhere, especially at night. And numbers, moreover, abound in collections of like things and in distances and sizes. All of these things have doubtless served to suggest the objects of mathematics.

"The point is that mathematics consisted, at the beginning, of abstractions of just such objects and phenomena. This was natural because there was so little abstracting to do. These objects, as held in the mind, were already rather abstract. We should not be surprised then, that abstractions spawn further abstractions, some of the latter being in the habit of cropping up in the real world. This taxi, from each minute to the next, expresses zook after zook in the orthogonal group of rotations about an inertial frame."

Although I wasn't quite sure what he meant, the taxi, without abandoning its inertial frame, took us out to the Oxford Ring Road, then off to a little country road that led to the village of Godstowe. There were thatched roofs, stone walls, and hearty-looking people with faces ruddy in the sunset. Peacocks strolled the gardens of the Trout. Inside, the evening was well under way. We were lucky to find a free table near the fireplace.

As we waited for service, Brainard expanded on his earlier views about mathematics and computers. "Whether or not the human brain is some kind of fancy computer, I have no idea. But there is evidence that we proceed in our researches at both a conscious and an unconscious level." He settled back in his chair to enjoy the fire, then withdrew his enormous pipe, which he lit with enthusiasm. "I have so few pleasures left," he explained. A cloud of aromatic smoke diffused gently into the surrounding haze.

"Take the case of Henri Poincaré, the renowned French mathematician at the turn of the past century. He had been working on a problem in the so-called Fuchsian functions—a tricky, deep, and difficult problem that he was having little success with. He was, he realized, overworking himself and badly needed a holiday. He was therefore grateful when the opportunity of an expedition came up. He went from Caen, where he lived, to Coutances, to meet up with friends. There he boarded a bus for the expedition. As he placed his foot on the first step of the bus, something strange happened. He had been discussing some matter with a companion, something entirely unrelated to the problem of the Fuchsian functions. But when he put his foot on that first step, the solution came to him in all its fullness. Making sure he could remember the main elements of the solution, he promptly dropped the matter and when he returned home later, he wrote out the details.

"It's a simple story, but one with important implications. Poincaré himself viewed mathematical work as creation. By this he meant that the mathematical mind, in solving a problem, worked by examining combinations of ideas. At first, the inquiry was conscious, but as the brain became used to the kinds of mental operations involved, it could, by degrees, take over the inquiry, or parts of it, at least. I mean that the mind could do this without Poincaré's direct knowledge for conscious participation.

"Poincaré likened his mental operations to little atoms that clung to the wall of his mind by hooks. Intense mathematical work had the effect of agitating these idea atoms, setting them in motion within the mind. In this way, new combinations of atoms could occur, the fruitful ones giving rise to new ideas that led to a solution. The analogy is mechanical, however, and runs somewhat counter to Poincaré's view of mathematics as created."

"Why, then, did Poincaré claim that mathematics is created?" I asked.

"Well, if you want the opinion of an old-fashioned Englishman, I would say it was a cultural thing. Perhaps discovery wasn't good enough for Poincaré. Being French, he would have to be a creator, an artiste of sorts.

"And to answer your unspoken question, I will have to say that my own mathematical work has always been conducted under the impression that I was discovering things. Things I had wanted desperately to be true have turned out not to be, and things that I could hardly imagine to be true have turned out to be. But do not ask me whether mathematics exists independently of the mind. I can only tell you that it exists independently *in* the mind. As for the power of mathematics to describe the physical world, the thing is a complete mystery to me."

Our steak and kidney pies arrived just then. Brainard immediately extinguished his pipe and set to work with the vigor of a 20-year-old. We talked little while we ate. We had denied ourselves for a long time.

When we were finished, Brainard promised that although he had some meetings to attend at his college the next morning, we would meet again in the afternoon, when he would explain his views about mathematical logic, mechanized proofs, computers, and what he called the "engines of thought."

CHAPTER 8

Mind Machines

"I've been thinking," Brainard began, but he stopped to fill his pipe. We were walking from the Churchill Hotel to the Mathematical Institute at the north end of Oxford. "I've been thinking about what yesterday I called the "engines of thought." I see plainly that, although it's not exactly what you had in mind with this *holos* business, you might achieve some harmonization of these two views of mathematical reality. But that is for later, perhaps." His pipe went out, and he applied another match.

"There are some items of unfinished business from yesterday, things I wanted to set right. When I spoke yesterday about the influence of mathematics on the world, I assumed you meant the direct influences or what you call the "unreasonable power of pure reason." There are the direct influences, of course. For example, there is a group called SU(3) that describes the quark configurations allowable in neutrons and protons. But there are also the indirect influences, and these, in a curious way, are just as important as the direct ones."

At last, his pipe began to belch clouds of smoke, and Brainard proceeded down High Street like an old-fashioned steam engine, I the caboose.

"The indirect influence of mathematics on the world proceeds by a circuitous route. Contrary to the wishes of my esteemed but disincarnate colleague G. H. Hardy, every bit of mathematics we discover becomes a potential tool for the description and understanding of the physical world, the material cosmos on which our existences allegedly depend. In this role, mathematics has always been, is, and always will be the primary source of precise abstract models for science. In this way, mathematics influences science. Now science, as you know, is the primary source of information and ideas for the development of new technology. Many, many scientific discoveries, from electricity to atomic fission, have resulted in new devices, from telephones to nuclear reactors. Now humans use all of these marvelous inventions and, in their myriad technological activities, humans affect the world profoundly. So there you have it: From mathematics to science to technology to people, we cannot doubt that mathematics has more influence on the affairs of humans than does any other field of human endeavor."

His pipe had gone out again, so we stopped. He had finished his take on the influence of mathematics on the world. I longed to ask him about the significance of the fact that all these machines worked, to tie it back to the real world in more direct fashion, but Brainard was clearly impatient. The pipe caught again, and we continued.

"The second item concerns meaning. Some people ask, 'What is the meaning of mathematics?' My reply is that mathematics by itself has no meaning, in the ordinary sense of the word. You could say that mathematics is all about horping zooks. Yet, if you consider that the meaning of a word always depends on a referent, then the meaning of mathematics would depend on its referents, the things to which it applies.

"Part of the meaning of a blorg—excuse me, a group—will be the real-world situations to which it applies. That little example of the rotations of a square I described yesterday could be taken as the meaning of the group whose table I showed you. So if we allow that the meaning of mathematics is clearly separated from its internal operations, we will proceed to step into a void of meaningless marks on paper. These considerations will lead, inexorably, to the engines of thought."

Brainard was on a roll. His pace was brisk and energetic, his thoughts flowed with the precision and power that recalled his heyday.

"All that off my chest, I want to show how mathematics as a meaningless subject became itself the object of a most amazing inquiry. This inquiry led to the view of mathematics as the accumulation of meaningless marks on paper. The manipulation of these marks was early recognized as a mechanical process. My story is about the manner in which mathematics became increasingly the province of machines that can think, after a fashion. The story begins with a field called metamathematics. You've heard of metamathematics, of course."

I had, but I sensibly refrained from saying that I had never met a mathematics I didn't like. "Of course," I responded, "but I'll admit that apart from a one-semester graduate course on computability, I —"

At that, Brainard launched into it: "We could define *metamathematics* as the mathematics of mathematics. Strangely enough, although metamathematics is about mathematics, it's also a part of mathematics. In any case, despite all the other mathematical advances of this century, I'm still inclined to award the palm to metamathematics—and to its subsequent influence. Bit of a mouthful that word, isn't it? Some people simply call it mathematical logic, a more descriptive term, if somewhat longer." His pipe caught at last, and we continued our walk.

"By the end of the nineteenth century, while most mathematicians motored on in much the way they always had, some—such as David Hilbert of Germany—had become concerned over the possibility of contradictions arising within mathematics itself. Alarm bells had been ringing since antiquity. You know of Zeno's paradox, I suppose."

I admitted I had heard of it: Achilles and a tortoise begin a race in which the tortoise is allowed a head start. Zeno argued that Achilles would never catch the tortoise, because for every place the tortoise passed, Achilles would have to pass that place before he could catch up to the tortoise. The problem was that Achilles would have to pass an infinite number of such places. For example, he would first pass a place halfway to the tortoise, then a place that was two-thirds of the way, a place three-quarters of the way, and so on forever.

"The problem in Zeno's paradox is that infinity can contain a subset that is just as large as infinity itself. Achilles was required to pass through an infinite number of places before catching the tortoise. Yet these places amounted to a mere subset of the places the tortoise has already passed. The seeming contradiction was even more apparent when you put the even integers in one-to-one correspondence with the integers as a whole."

Brainard stopped to retrieve his notepad, writing the following sequence, which I could barely see because the page fluttered in a sudden breeze.

1	2	3	4	5	6	7	8	9	10	11 . . .
2	4	6	8	10	12	14	16	18	20	22 . . .

"How could the concept of infinity ever become as well-defined and respectable as that of an ordinary number when such anomalous phenomena loomed over it?"

The general drift of Brainard's remarks was making me a bit apprehensive. Were these contradictions and anomalous phenomena about to turn nasty? I wondered. I therefore greeted Brainard's next remarks with some relief.

"It was Georg Cantor, a Russian-German mathematician, who resolved these seeming paradoxes that circled the notion of infinity. It was a brilliant piece of work. He defined infinity through infinite collections. Such collections were characterized by the fact that you could subtract a finite number of elements from them without changing the size of the collection. You could even subtract an infinite number, as in the example of the integers.

"But some mathematicians, such as Hilbert, were ever watchful in the matter of possible contradictions arising in their field. Hilbert decided that the best way to ensure that the main jewels of mathematics—such as set theory, arithmetic, and the calculus—were immune to such difficulties was to recast these fields in the language of logic. Mathematics was about to become mere 'marks on paper' with a vengeance.

"Hilbert proposed that standard theories be proved free of contradictions once and for all, by the simple expedient of rederiving them in a content-free manner. He showed how to express mathematical theories in terms of sequences of symbols called *formulas*. The axioms on which a theory was based could themselves be cast as formulas, as could all the theorems forming the theory. Now to derive that theory in a content-free way, you will notice that Hilbert paid no regard to the actual meaning of the terms, variables, and other objects within the theory in their new incarnations as mere marks on paper.

"You see, that's what I was getting at yesterday with the talk of horping zooks. Good gracious! You'd think I had gone mad! Hilbert showed how to translate all of mathematics into formulae that could be derived via simple rules of logic, from other formulae. Mathematics was reduced, in effect, to the mechanical manipulation of meaningless marks on paper."

"Strange," I interjected, "that's how most people think of mathematics anyway."

"Quite!" replied Brainard with a hearty laugh. "Of course, when we say meaningless, we're only referring to the kind of external meanings I mentioned at the beginning. You know, the instances, applications, and what have you. But mathematics has an internal meaning, as well, this being its logical structure, in essence."

The talk of contradictions had still left me a little anxious. "Tell me," I asked, "how Hilbert's program for mechanizing mathematics would free it from contradictions."

"Hilbert hoped that once he had cast a mathematical theory in his special, metamathematical formulation, he would lay bare the structure of the theory itself, drained of referents, of associations, of hidden ideas, all the things that might subvert the proof process. He had in mind what he called a *beweistheorie,* a theory of proof that could be applied to the structural features of a given mathematic. The project would be immense, well beyond the capabilities of a single person.

"Now a few decades later, the British philosopher-mathematicians Alfred North Whitehead and Bertrand Russell published their *Principia Mathematica,* an attempt to axiomize all of mathematics, to put the whole edifice of modern mathematics, at least the basics, on a thoroughly solid footing. There would be no logical holes, every proof complete, every new theorem built firmly on the foregoing. Of course, it was not quite the *beweistheorie* that Hilbert had in mind, but it was a stunning achievement, really."

The possibility of internal contradictions that Brainard had just raised was a serious matter, indeed, especially for someone like me, a quasi believer in the holos. If there were contradictions, I reflected, the whole idea of the holos would come crashing down around my ears, an empty concept from the center of which I would hear the death groans of Pythagoras, the cries of the Brethren of Purity, Kepler, Balmer, and the others in anguish of soul.

"I suppose that Russell and Whitehead did not prove the consistency of mathematics," I ventured to conjecture.

"Insofar as their effort did not amount to the *beweistheorie* that Hilbert had in mind, the answer is no. And Russell and Whitehead knew it. But their ultimate dream of a final proof of consistency was absolutely shattered by a most upsetting paper published just one year after the *Principia* appeared. The paper was written by a young German logician and mathematician named Kurt Gödel. In it, Gödel proved his famous incompleteness theorem. The theorem showed that any mathematical system containing the standard arithmetic of the integers will be either inconsistent—that is, will contain contradictions—or incomplete. The latter word [*incomplete*] means that there will be *theorems*—true statements within the system—that are nevertheless simply not provable in the system. The price of consistency, according to Gödel, is incompleteness. Strange, isn't it? It still strikes me as strange today.

"Some philosophers have made much of Gödel's theorem, too much in my view, by declaring the superiority of human reason over mere mechanical reasoning. It was a human being, after all, who had pointed out the impossibility of proving certain theorems by the kind of mechanical reasoning that Hilbert had proposed. They quite forget, however, that the metamathematical process itself could be made mechanical in the same way as the Hilbert program. Indeed, we are faced with an endless regress of systems. System upon system."

Brainard sighed, as though carrying the burden of all those systems in his mind. I tried to lighten his load: "Has any mathematical system been proved consistent?" I asked. At this question, Brainard stopped on the pavement to stare at a distant spire. Then he withdrew his notepad, scribbling as he talked.

"My word, yes. There's the propositional calculus, after all. An almost trivial observation makes the matter easy. As you know, the propositional calculus is the simplest form of logic. It consists of *propositions,* statements that can be either true or false. We symbolize these propositions by letter names such as p, q, r, and so on, the so-called atomic propositions. In the propositional calculus, we can make slightly more complicated propositions out of atomic propositions by combining them in various ways. Thus, 'p and q' is written $p \wedge q$, 'p or q' is written $p \vee q$, and 'p implies q' is written $p \rightarrow q$. And don't let me forget negation, not p, written $\sim p$.

"The meanings of the simple expressions I have just mentioned are only determined in relation to truth. In this sense, their precise meanings are pretty much as they sound. For example, $p \wedge q$ is true if both p and q are true, whereas $p \vee q$ is true if either p or q or both are true. The expression $p \rightarrow q$ means that if p is true, then q must also be true. Finally, we have $\sim p$, the negation of p. If p is true, then $\sim p$ is false, and conversely.

"There are simple rules for building up expressions in the propsitional calculus. In brief, you can take any two propositions, no matter how simple or complicated, and join them together with the and symbol, $\wedge$. You can do the same thing with the or symbol, $\vee$, or the implication symbol, $\rightarrow$, and you can negate a proposition, no matter how simple or complicated, by simply putting a negation sign, $\sim$, in front of it. Here's a more or less typical expression in the propositional calculus, for example:

$$(p \wedge q) \vee (\sim(p \wedge q) \rightarrow q)$$

"Now this particular expression can be either true or false, depending on the truth values of the participating expressions. These are the theorems of our subject. The axioms from which the predicate calculus is derived have a particularly simple form. Here, for example, are the axioms used by Russell and Whitehead in the *Principia:*

1. $(p \vee p) \rightarrow p$
2. $p \rightarrow (p \vee q)$
3. $(p \vee q) \rightarrow (q \vee p)$
4. $(p \rightarrow q) \rightarrow [(p \vee r) \rightarrow (q \vee r)]$

"Now you can start from these axioms and use two rules—called substitution and detachment—which I will describe in a moment. Building out from the axioms, guided by intuition or not, you will arrive at a succession of propositions that are universally true in the sense I have just defined. For example, here is a theorem I could prove in this manner:

$$p \rightarrow (\sim p \rightarrow q)$$

"No matter whether p is true or false or whether q is true or false, this statement always evaluates to 'true.' This particular theorem, although fairly typical for the shorter theorems, has great significance for the system as a whole.

"You see, if the propositional calculus is consistent, then it should never be possible to derive a theorem T along with its negation $\sim$T. Suppose we found such a disastrous proposition T. We get nowhere until the fruitful idea descends like the apple upon Newton. What would happen if you substituted the frightful theorem/antitheorem T for p in the preceding little theorem? It's the rule of substitution that allows us to do this:

$$T \rightarrow (\sim T \rightarrow q)$$

"I have substituted the proposition T for the symbol p in the foregoing theorem. The other rule, the rule of detachment, allows us to detach any expression that is implied by an expression that we already know to be true. Because T is allegedly true, we may detach $\sim T \rightarrow q$ from the formula, finding that $\sim T \rightarrow q$ is universally true, like all the other formulas generated by the system. But wait! What's this? Because T is self-contradictory, not only T is true, but $\sim T$ as well. This enables us to apply the rule of detachment once again, detaching q as a universal truth from the formula $\sim T \rightarrow q$. We obtain in the end the simplest possible theorem, q.

"But wait again! Because by this line of reasoning, q is itself universally true, it can be replaced by any propositional formula whatever, according to the rule of substitution. Odds bodkins! Everything is true! If the propositional calculus contains a contradiction, then all propositions, no matter how complicated, are also true—all propositions."

Brainard emphasized the word *all,* giving me once again that peculiarly intense stare.

"And now, once again, comes the crunch: Are all propositions in the predicate calculus also theorems? By no means! For example, the proposition $p \vee q$ is not a theorem. When p and q are both false, for example, so is this proposition. Thus, by stepping outside the propositional calculus and into the metamathematical level, so to speak, we have found a contradiction to the assumption that the propositional calculus is either inconsistent or free of contradictions. We may also prove, by the way, that the propositional calculus is also complete. Any theorem of propositional calculus may be derived from the axioms.

"More powerful mathematical systems are provably not in such a simple case. As Gödel showed, any mathematical system that contains the arithmetic, if consistent, must be incomplete. All such mathematical systems, and that includes most of the really interesting mathematics, are marked by occasional signposts that read, 'YOU CAN'T GET THERE FROM HERE.'"

I asked Brainard whether he knew of any theorems that are not provable in any of our standard mathematical systems.

"I'm sure you mean *suspected theorems.* At one time, some wondered whether the famous four-color conjecture might be in this case. You know—the idea that you need no more than four colors to distinguish the countries of any map, no matter how abstract or unreal. Before this conjecture was finally proved in the 1970s, and only with the help of a computer at that, some thought it might be one of the strange theorems hinted at by Gödel's theorem. At present, one of the few candidates is *Goldbach's conjecture,* which states that any even number is the sum of two primes.

"Here, try it for yourself. What about 28?"

"Umm, 1 plus 27 no, 2 plus 26 no, 3 plus 25 no, 4 plus 24 no, 5 plus 23 yes. Yes!" I blurted out.

"It would appear that no matter what even number you start with," Brainard continued, "you will always be able to find two prime numbers that sum to the even number. It's so simple you could explain it to the proverbial man on the street, yet no one's been able to prove it!"

Brainard stopped to gaze abstractly into the window of a computer store. "Now we have talked about metamathematics and the problem of proving mathematics consistent, but I've said nothing about a most important direction in which these ideas led. The computers in this window remind me of a curious little method called the British Museum algorithm. It may be something you're looking for, as it illustrates the notion of the independence of mathematics from cultural influences, as regards the central truths or theorems of these systems. The point, the ultimate point of the marks-on-paper view of mathematics, as encouraged by metamathematics, is that you don't even need humans to find new theorems. There are machines that are perfectly capable, at least in principle, of discovering new theorems.

"A computer, programmed with the British Museum algorithm [BMA], could begin with the Russell–Whitehead axioms, for example, and systematically generate all possible theorems by applying the rules of substitution and detachment to the axioms, to obtain the first layer of theorems. The fact that most of these theorems are trivial matters not. They are fodder for the next level of processing, wherein the BMA computer uses the same rules to generate a new layer of theorems. Before long, the computer will be generating some genuinely interesting theorems. It must, because sooner or later, it will get every theorem."

"I understand from my colleagues in computer science," Brainard continued, "that such an engine could be built, at least in principle. In fact, I believe some of the early artificial-intelligence researchers in America actually built a version of this program, and it included a rule for getting rid of some of the less interesting theorems, mechanically."

I happened to know a little about the research Brainard had mentioned, and he appeared to have taken the idea a little farther than it would stretch. "But that project used only the propositional calculus, didn't it?" I asked.

"That one, yes. But other mathematical systems are also in principle mechanizable. All you need are the axioms and the rules of deduction for the system in question. That may be a point you want to explore. For example, I can readily visualize a room in which several mathematicians sit with pen and paper. Another person sits before a computer programmed with the British Museum algorithm. Of course, we will assume that the program is fast enough to compete with the humans. In any event, I can see one human after another stopping to write down a new theorem. And about as often, the person running the computer writes down a new theorem, too.

"Well, when you examine the list of theorems that the mathematicians have found, whether they be from Turkey or Tibet, you will find a large number that are identical or intimately related, mathematically. Moreover, every theorem the mathematicians find will be found, sooner or later, by the computer. In this milieu, independent discovery is hardly a mystery, and without committing myself to the view you seem to hold, I can certainly say that it's a mistake to regard independent discovery as a mere coincidence, whether mechanical or cultural. What the computer example shows is that the theorems, whatever their status in your holos, are certainly implicit in the system. They await discovery, just as the number 37 reasonably awaits recitation when a child counts to 100."

We had gravitated across the street, for some reason, and now stood in front of a famous old church, St. Mary Magdelene. Ever since leaving the hotel where Brainard had mentioned the engines of thought, I had been waiting for him to broach the subject directly. Impatient, I asked, "Are the engines of thought just computers, then?"

Brainard pointed to a low, two-story building just down the street. "I will have more to say about it when we get in there."

We entered the Mathematical Institute to find ourselves in an ample foyer and a large tearoom from which came a burst of loud laughter. Brainard showed me some of the pictures on the wall.

"Ahh. Now this is G. H. Hardy, a true Oxonian and one of the best mathematicians of the early twentieth century. It was Hardy who said there is a mathematical reality outside of us. He went even further, declaring that it forms part of physical reality, without saying in what sense it is a part."

A knot of people had gathered behind us.

"Good afternoon, John," said one of them. "You're just in time for tea, then. How have you been keeping?"

Faculty and graduate students were in this bunch. Evidently, Brainard did not come to the institute often, owing no doubt to his advanced age.

"I am being pushed to the very limits of my endurance by this Dewdney chap," he replied. "He has been plying me with questions of mathematical philosophy to the point where my brain is about to burst!"

He introduced me around, and we adjourned to the tearoom. We sat down at a long table, and, after much rattling of cups on saucers and adjustment of cakes on plates, Brainard introduced me to the others at the table, mentioning that I had been converted by some Greek to the novel idea that all mathematics exists in a place called the *holos*. This introduction left me feeling somewhat uneasy, but I could see from many smiles that everyone assumed Brainard was having them on. Then he simply picked up our previous conversation where he had left off.

"I was about to explain to my guest some curious ideas about the theory of computation. It's speculative, I'll admit, but there are some interesting quasi questions attaching thereby."

One of the graduate students wondered what a quasi question was, and, to a burst of laughter, Brainard declared it to be a question that has a quasi answer. He was in his element, lecturing once again.

"Let us begin a sort of game," Brainard continued, "where we imagine a computer constructed by a race of alien beings on some other planet of our universe—or in some other universe, if you like. It makes no difference. My question is this: What sort of computer might it be? What functions might it compute, what mathematical principles might it embody? The quasi answer is that whatever it might do, I believe it would be incapable, in principle, of computing anything that our computers could not. I wonder who knows why I believe this?"

"Because you're a doddering old fool," remarked a professor across the table jovially. It was Weisskopf, a senior mathematician at the institute, who was known for his outrageous sense of humor. Another burst of laughter rolled forth as Brainard smiled.

"Right on all three counts," he murmured contentedly, then looked around the group. "Anyone else?"

"I feel sure you want someone to say 'Church's thesis,'" a graduate student piped up.

"Church's thesis it is. As everyone here undoubtedly knows, Church's thesis says that all computers are created equal, in a certain precise sense. It really is a startling proposition, when you think about it. Everyone believes it to be true, yet no one has the slightest idea of how to go about proving it. Perhaps it's one of those Gödelian, you-can't-get-there-from-here propositions.

"By the way, not to ramble on, but I wonder how many mathematicians are as pleased as I am by the prospect of theorems that we shall never prove. Perhaps they are all off in some dark corner of Dewdney's holos, but I would like to have one near to hand, to know it before I die. What scrambling there will be, what probing and what new mathematics spawned! Where was I?"

"Church's thesis," prompted a student.

"Of course. Church's thesis arose from a very peculiar circumstance during the early history of the theory of computing. By the early thirties of this century, no less than three completely different abstract models of what it meant to compute a function had been proposed. As I hope you realize, there were no computers to speak of anywhere at this time. Computers were only a gleam in the eye of a handful of mathematicians, including Alan Turing, Alonzo Church—the American logician—and a few others. The idea was in the air, so to speak—a kind of zeitgeist.

"Church had formulated a new way to compute things, called the lambda calculus. But another, seemingly quite different, way to define computing had already been published. It was called the theory of general recursion. Church was astonished, when he compared the lambda calculus and general recursion, to discover that they computed precisely the same things. Now this may not be immediately evident to some of you, but in the most general sense, when you set out to define a system for computing functions, you should expect to find that some functions are computable by your system, and some are not. Yet Church found that the two formulations, in spite of their wide differences, were nevertheless completely equivalent in this sense. They computed exactly the same functions, albeit differently. He even tried to think of a way to compute that was not equivalent to these formulations but could not. This led him to declare, rather too boldly in the opinion of some, that any scheme proposed now or in the future would turn out to have the equivalent computing power of general recursion of the lambda calculus. We call it Church's thesis, something a bit different from a conjecture because it was not all that well defined. In any event, it is not called Church's theorem. It may not be a theorem, of course. There may be a much more powerful notion of computability, but we have good reasons to doubt it.

"You see, very shortly after Church stated his thesis—that all computers are, in a precise sense, created equal—along came a paper by Alan Turing, in which the Turing-machine notion of computing was first described. In that paper, Turing showed that the Turing machine (he himself did not call it that, of course) was equivalent to both the lambda calculus and to general recursion. Again, Turing machines computed exactly the same functions as the other two schemes. Moreover, Turing's formulation of what it meant to compute was much different from the other two schemes—more different than they were from each other!

"Since that time, literally dozens of different computing schemes have been proposed. As long as they involved a finite set of definite rules for manipulating a finite set of symbols, they turned out to be equivalent to all of their predecessors or, occasionally, fell very short indeed. I say, there is very heavy game afoot here, mathematically speaking. If Church's thesis is true and an alien being constructs a computer satisfying these minimal constraints, that computer will be not one whit more powerful than our own. There would be no function it could compute that our computers couldn't. It might be faster, of course, or even slower, but essentially no different.

"Now the Turing machine is a very strange sort of machine. It is abstract, of course, but you will notice that it is a very action-oriented sort of thing. It reads symbols from a tape and, in response to those symbols, writes other symbols. It is controlled by an internal table that tells it, for each symbol it might encounter on its tape, what symbol to write in its place and which way to move the tape for the next cycle of operation. Shall I draw a picture?" He removed the infamous notebook from his pocket.

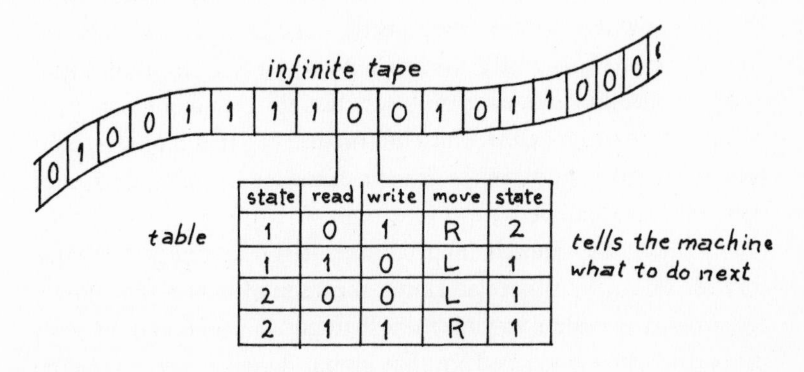

state	read	write	move	state
1	0	1	R	2
1	1	0	L	1
2	0	0	L	1
2	1	1	R	1

A Turing Machine

"Now for my purposes, it's not important to grasp precisely how a Turing machine operates. But it is important to notice that when you change the internal table, you change the machine. There is another kind of machine, called the universal Turing machine, which is supplied with a fixed table and an additional program tape that it may consult as it computes on its main tape. Each time it encounters a new symbol on its main tape, its table causes it to consult the program tape about what to do. This universal machine is just an abstract version of the modern digital computer. It stands for all the digital computers on Earth, although none of them resembles Turing machines. If Church was right, then the universal machine stands also for all possible digital computers everywhere in the cosmos—past, present, and future.

"Now our alien friend—if he, she, or it probed the matter a little—would, I think, run straight into Church's thesis, calling it perhaps Blorg's thesis, since he wouldn't have the slightest idea who Church was." Brainard's dry delivery brought more chuckles.

"The point of this observation is to answer a question posed by my guest. Dewdney has asked whether I can provide any evidence for the independent existence of mathematics or explain its unreasonable powers.

"Computers establish at least one thing about mathematical realities. How can anyone doubt the independent existence of the numbers 0 and 1? They manifest in many different forms in a computer. They are now either the point of light on the screen or its absence, now the charge in a transistor or its absence, now the presence of a pulse in a wire or its absence, now a '1' printed on paper or a '0.' They flit from one form to another, forever aloof from final definition but undeniably real. They no longer depend on the mind of man, and the evanescence of their existence is the very proof of their ultimate reality.

"Precisely the same problem concerns the reality of genes, those tiny bits of protein that determine so much of who and what we are. Are genes real? At death, these tiny bits of protein all die with us, and yet they go on and on through our descendants. The information in a gene may be transmitted from generation to generation for thousands, even millions, of years, without change. What then is the gene? It is not just the protein but the pattern it embodies, a completely mathematical concept—nothing more, nothing less. We may express that pattern with a four-letter alphabet, but the expressions are all arbitrary. In a certain crucial sense, the pattern of the gene is more real than the particular expressions it finds in bits of proteins, even though, like the numbers in a computer, it depends on the physical substrate to keep it alive, so to speak.

"Once you have admitted this determinative reality 0 and 1, can the remaining integers be far behind? And what of the formulae and expressions in which they participate? Can they be any less real?"

There was a lengthy silence during which Brainard swallowed frequently and turned a little pale.

"I say, are you all right?" It was Weisskopf once again.

"I am dreadfully tired. Would one of you be a good chap and run Dewdney out to Whytham Abbey? David Gridbourne keeps his computer out there, you know."

Brainard had mentioned that I might enjoy meeting Gridbourne, who had a reputation as something of a mad scientist. Gridbourne, Brainard informed me, thinks he has created living creatures in his computer.

Before I left, I thanked Brainard profusely. He took my hand almost convulsively. "For God's sake, treat this subject with the care that it deserves, and above all watch out for Gridbourne. Terribly nice chap but actually a lunatic!"

A young mathematician by the name of Winslow drove me out of Oxford, onto a country road, and into the village of Whytham, with the obligatory stone walls and thatched roofs. I asked him about Gridbourne. Winslow knew him vaguely and confessed to being curious about the "Whytham Wonder," as he was called around Oxford. Near the center of the village, Winslow made a sharp turn into a courtyard, grinding to a halt before a stone structure of indeterminate age. We entered through great oak doors and found ourselves in a court lined with doors to various flats.

"The lab's over here, I believe." Winslow strode toward a door.

The door opened to reveal a harried-looking man of middle

age and steel-gray hair, great fleshy lips, and a deep voice.

"Brainard thought you might be willing to entertain a visitor, even if briefly," Winslow explained somewhat nervously to Gridbourne, who peered curiously past Winslow.

I suppose you want to see the bloody two-dimensional universe," Gridbourne snapped at me.

He seemed in a nasty mood, and I hung back until Winslow had introduced us. We entered a palatial room that led out to a patio with a rose garden behind it. Gridbourne, who was a member of the institute, rarely visited his office, so immersed was he in his computing project. Having independent means, as the British called it, he was free of many strictures, including the necessity of applying for grants. We examined the machine, a set of powerful Sun computers with massive disk drives, all in a bank against one wall.

"I never turn this machine off. It has a backup generator in case of power outages. For the past two and a half years, I have been running a simulation program called 2DWORLD, essentially a cellular automaton. Imagine a two-dimensional universe in the shape of a vast sphere. It is divided up into tiny square cells, and what happens in each cell is determined by a few simple equations that mimic to a certain degree the equations of modern physics.

"I began with a state in which each cell of this universe was assigned a 0 or a 1 at random. I wasn't terribly happy with the rules and had meant to debug them with a few test runs before I started probing these axioms in earnest. However, what happened in the first hours, then in the first days and weeks, absolutely prevented me from ever turning the machine off or allowing it to be turned off."

"What did you see?" I asked, becoming intrigued.

By way of an answer, Gridbourne pressed a key and the screen came alive with strange patterns of little squares. The patterns expanded and contracted in a regular manner, exchanging small bursts of bright dots, like the lightbulbs on a Times Square display. Gridbourne pressed another key, and the scale of the display changed, as though a camera were backing away from the scene, showing us a larger picture.

"What you just saw was the fundamental physics of this

two-dimensional cosmos. Now we're looking at a rather large molecule that has become increasingly numerous in my little cosmos since last Christmas."

Indeed, a large, complicated structure now wobbled before our eyes. Gridbourne pressed the key again to reveal a new level of integration in which molecules—some large, some small—circulated within a crude, circular membrane.

"Is that what I think it is?" I asked.

"What do you think it is?" he asked, looking up incredulously.

"Well, I must say it looks alive, somehow."

"It is," he responded contentedly. "Certain rules of physics seem to guarantee the emergence of levels of organization, one after another, without apparently ending. At this level, the system has arrived at structures that propagate themselves endlessly. And they've been changing since they first emerged. They're definitely getting more complicated, and they have a genetic code of sorts, although based on very different structures from our own."

"Are they in fact evolving?" I asked.

Gridbourne looked quite pleased with this question. "I don't know. I suppose that is the question that keeps me suspended in front of the program. As I am sure you're aware, an indefinite evolutionary scenario may lead, ultimately, to intelligent creatures, albeit two-dimensional ones. That would put me in a pretty pickle indeed. On the one hand, it would be the scientific feat of the century, not to say the millennium. On the other, I would feel responsible for these creatures but completely unsure of what to do."

After some reflection, I asked the question closest to my heart. "They might develop science, I suppose, discovering the laws that you have installed in their cellular space. Do you think they might ever discover that they exist only in a computer?"

"Perhaps. But they would have no idea what computer or where. Computers have a thousand possible homes, each based on an entirely different technology. In principle, you could build a computer out of ropes and pulleys, a really massive, very slow affair covering thousands of square miles, and in this computer, you could have the equivalent 2DWORLD program running. And the creatures would have no idea that their space was an enormous raft of zeros and ones stored in a rope-and-pulley computer, or for that matter one made of bamboo and silk, or electrons in silicon, or water jets in plastic channels, or light in fiber optics, or what have you. It would be a barrier that they could never penetrate.

"By the way, I'm not really mad. I don't believe for a second that we are in the same fix as any emergent creatures in my system. But you have hit upon a most troubling point, one that lurks around the foundations of any inquiry into the necessary limits to our knowledge. For behind those limits, there may be truths too deep to bear, including the reasons for our existence."

Gridbourne was a man immersed in a phenomenon, and this last comment seemed to set his mind back to the work at hand. Though he had turned quite cordial, he now tensed up again and hurriedly ushered me from his laboratory.

When I returned to Oxford, Brainard had left for his house to nap the rest of the afternoon, so I did not have a chance to reflect with him about Gridbourne's work. My train left Oxford station at 5 P.M. With only the echoing horn and rumbling wheels to accompany my thoughts, I vaguely watched a replay of yesterday's scenery while my own engine of thought turned the questions over and over again. I wondered why Brainard had sent me out to the abbey. Was it to witness the ultimate act of an engine of thought—namely, to become the home for a miniature universe? Was this his sly hint that the holos was real after all and living in some behind-the-scenes cosmos computer? If so, my quest was doomed without necessarily being wrong. Instead, was Gridbourne's two-dimensional world only meant to demonstrate that computers, sharing with humans a certain facility for manipulating symbols, represent the ultimate expression of the independence of mathematics?

My mathematical adventure was over. It remained only to reflect on everything I had learned during my travels and to try to come to some conclusion. In the meantime, it seemed, I had relived the traditional practice of traveling in search of knowledge. I felt just a bit like a modern-day Thales; or like Fibonacci, bringing intellectual riches from four distinct corners of the world.

Epilogue: Cosmos and Holos

I was halfway across the Atlantic Ocean, enjoying the comfort of a rare indulgence (a first-class seat) when, quite suddenly, doubts about the power and independent existence of mathematics began to assail me. At that very moment, the aircraft began to shudder. The thrust of the engines no longer equaled the rearward drag, the airflow over the wings stopped obeying the equations of fluid dynamics, the strength of the airframe ceased to be proportional to the cross sections of its members. The plane broke apart, the fragments falling at arbitrary speeds and accelerations, some upward. I was lost.

Instead of plunging into the icy seas below, however, never to be heard from again, I remained in my seat. Then I opened my eyes. Everything was normal. The steward was passing out suppers, and no one seemed in the least alarmed. My bad dream had one good effect, however. It made me wonder what the cosmos would be like if it did not obey mathematical laws. Was such a thing conceivable? Looking at matters from the other side, one could also ask,

What would the cosmos be like if it did obey mathematical laws? How would we know the difference? I could imagine a whole research program aimed at this question.

At home once again, there was time to reread my voluminous notes and to transcribe the tape recordings of my conversations with Pygonopolis, al-Flayli, Canzoni, and Brainard. The time has come to summarize my thoughts and reflections on what I learned during my travels. Although I have tried to be objective, I cannot help but think that a holos of some kind lurks behind the cosmos.

During my travels, I had spoken with four scholars, not all of them eminent, to be sure, but each dedicated in his or her own way to a belief that something is going on. All shared a conviction that there is a profound connection between mathematics and the cosmos or, if you prefer, between holos and cosmos. I had listened intently to these earnest researchers in offices, courtyards, city streets, and ancient temples, over deserts and desserts. I had watched them make diagrams in the dirt, on chalkboards, in notebooks, on dinner napkins, and even in the night sky. If my own conclusions seem to lean heavily in their direction, I can hardly be blamed.

The central historical character of this mathematical minisaga is Pythagoras. We found Pythagoras's influence recurring throughout the history of mathematics and spanning many cultures. We saw evidence that the famed Pythagorean brotherhood had transmuted into the Brethren of Purity during the Islamic period and then was heard of no more. Pythagoreans continue to pop up through that same history, however. Kepler and Balmer, one suspects, are just the tip of an iceberg. Mathematicians with an interest in physics and a conviction that the role of mathematics in the cosmos was no accident would probably find the Pythagorean spirit congenial.

The Pythagorean theorem, with which this book begins, also pops up in more than one application: spatial distances, trigonometry, relativity, and so on. It appears in literally thousands of applications, probably more than any other theorem of mathematics. It does this not because of its antiquity, for we have hundreds of theorems from antiquity, all no less true today than they were then, but not all as useful as the Pythagorean theorem.

Let me begin by rephrasing Pythagoras's convictions about the ultimate structure of the cosmos, softening it and making it more precise:

The Pythagorean Hypothesis

The cosmos and everything in it is ruled by mathematical laws.

The hypothesis says nothing about how the cosmos came into being or why it has this extraordinary property, only that as we find it today (and as Pythagoras found it yesterday), there is nothing in the cosmos—no corner, no tiny part, no accident, no substance—that does not follow one kind of mathematical rule or another.

At the beginning of my journey, I was determined to ask two questions of my four consultants, and I remained faithful to this plan, sometimes receiving surprising answers. The questions were

1. Why is mathematics so incredibly useful in the natural sciences?
2. Is mathematics discovered, or is it created?

If the Pythagorean hypothesis is correct, we immediately have an answer to the first question: Suppose for a moment that the cosmos, including the Earth and everything on it, is in some precise sense determined by mathematical laws. Then mathematics is incredibly useful in the natural sciences because the job of these sciences is to uncover structure, and that structure just happens to be mathematical. The laws of physics, astronomy, and chemistry *must* therefore take a mathematical form.

Now it's one thing to uncover the structure of the cosmos, finding mathematics everywhere, but it's quite another thing to come upon mathematics without even considering what the structure of the cosmos may be. For this is how most mathematics actually *developed* (this word does not commit us either to discovery or to creation). Yet, if the cosmos has a mathematical structure (remember, we are assuming the Pythagorean hypothesis), then presumably, it had this structure from the beginning. The cosmos was apparently here long before human beings were, and so, therefore, was its mathematical structure. On this score alone, and in spite of some holes in my argument (which may or may not be filled in later), the Pythagorean hypothesis implies the preexistence of mathematics in at least this sense. We can answer the second question, "Discovered, most likely."

We can also take a look at the hard form of the Pythagorean hypothesis, in which the cosmos comes equipped with a holos, a place where mathematics may be said to have its independent existence, although that place is not necessarily in the cosmos. Given the existence of a holos, the second question becomes tautological: "Discovered, of course!" After all, asserting the existence of a holos is pretty much equivalent to saying that mathematics preexists. To my overgenerous mind, preexistence implies something waiting to be discovered.

At the other extreme, we confront a cosmos that is forever unknowable, a cosmos that has been wedged into culturally determined but otherwise arbitrary forms by a science that deludes itself about the absolute nature of reality. The leveling scythe of social constructionism imposes a rigorous democracy on thought itself. By fiat, no description of the cosmos can be preferred over any other. We must, by contrast, frame an alternative hypothesis in spite of the fact that social constructionism does not recognize hypotheses or their tests to be any more valid a route to knowledge than tarot cards.

THE POSTMODERN HYPOTHESIS

The cosmos, whatever that is, has no preferred description.

This position is alarmingly easy to defend, whether you know a lot of science or only a little. No proof, no evidence of any kind can hope to elevate the mathematical description of the cosmos to an inherently preferred, absolute, or special position. By definition, all descriptions are equally privileged. As every mathematician knows, you can't argue with a definition.

This view, which seems to require a special discipline to maintain, has its origin in the theories of the philosopher Thomas Kuhn, who argued that scientific revolutions are driven by culture or by changes in culture. Kuhn described "paradigm shifts," such as the Copernican revolution, as primary cultural events, signals of change in the way people understood the world around them. The case is well made, but entirely on a cultural level, and without giving the scientific conception of truth any special role to play.

Read closely, Kuhn actually only says that in the Copernican paradigm shift, the direction of astronomical inquiry changed after the dethronement of the Earth from its central position in the cosmos. This is perfectly true. Ironically, the Copernican paradigm replaced an earlier one that was apparently culturally independent! As al-Flayli was at pains to point out during that memorable night on the desert, the astronomers of ancient Egypt, Babylon, India, Greece, and Arabia all saw the sky the same way, as a hemisphere. Now their astronomical descendants "see" the night sky quite otherwise.

My adventures in Miletus, Aqaba, Venice, and Oxford (not to mention some subsequent reading and consultation) convinced me that while the form and even the direction of mathematical inquiries were driven by culture (in some cases, anyway), the results of these inquiries were not. How else can we explain why the Pythagorean theorem leapfrogged from culture to culture, why all theorems, no matter when they were discovered, have also done this? Also, when a theorem fails to vault through history, it gets rediscovered anyway! There are many examples of this phenomenon, including the theorem of ibn Qurra, as rediscovered by Pierre de Fermat. I am inclined to borrow the phrase of that tireless phrasemonger, Pygonopolis: Mathematics, like the wheel, is transcultural.

For example, Pythagoras inquired into the problem of commensurability by using visual diagrams for geometrical and numerical objects and applying logical arguments to them. Modern mathematicians prove the incommensurability of the side of a square with its diagonal by symbolic algebra and nothing very sophisticated, at that. The theorem remains unchanged: The side of a square has no common measure with its diagonal. Again, the Arab mathematicians of 1,000 years ago argued their algebra in words, giving it a completely different appearance from modern algebra, but the content is the same. Even at the level of individual concepts, this rule seems to hold. Number itself is transcultural, as al-Flayli pointed out: The Arab shepherd sold 42 sheep to the Byzantine merchant, who was quite satisfied with XLII of them.

In short, I found no support for the postmodern hypothesis, except in the vastly curtailed form with which Kuhn first proposed it, a form with which no serious scientist will argue. Of course, the bold hypothesis framed here is by definition nonfalsifiable and therefore lies beyond the reach of reasoned argument.

Maria Canzoni's charming twist on the fable of the Blind Men and the Elephant provides an alternative view of paradigm shifts. Take the wise man who first examines the elephant's foot. Feeling the great toenails, he declares, "All bodies attract each other according to the inverse square of their distance apart." This is Newton's law of universal gravitation. Later, another wise man feels the extent of leg above the foot. He declares, "The presence of matter distorts space-time in a manner that creates an attraction between two such distortions." Newtonian physics, as Canzoni was at pains to point out, is the foot at the end of the leg, a special case of Einstein's more general theory.

To the extent that this analogy is valid, the term *paradigm shift* contributes little, if anything, to understanding the elephant. In fact, it tends to retard understanding. As far as mathematics is concerned, the hypothesis that mathematics amounts to little more than cultural meanderings can only be defended by ignoring the evidence. There is, as Sir John Brainard might say, "very heavy game afoot here." It must be Canzoni's invisible elephant.

On reflection, one central phenomenon recurred throughout my adventures in the four corners of the world, a phenomenon that I did not anticipate at the beginning of my journey but that is now patently clear. It is captured by the phrase "essential content." Every mathematical idea, from the concept of number to the most sophisticated theorems, has an essential content that defies every attempt to describe it in a way that does not amount to yet one more expression of that content, a phenomenon that eerily echoes the social constructionist view of things. No expression is preferred.

What is the essential content of the number 42? It is not "42" or "XLII," nor is it "101010." It is not "**," nor is it 6 times 7. Yet the essential content is expressed by each of these means, rightly understood. Essential content retreats before every attempt to define it, like the Zen koan: not this, not that. What is the essential content of a circle? It is not any of the infinity of circles that we may draw or algebraic formulas for circles that we may write. What is the essential content of the Pythagorean theorem? We can state the theorem in English or in ancient Greek. We can represent it by a diagram or by an algebraic equation, but it is none of these things.

Yet essential content is a perfectly real thing, as Brainard pointed out with his computer example. The concepts of 0 and 1, apart from the numerical symbols just written, appear in a computer as patterns of voltage in electronic registers, as points of light or dark on a display screen, as pulses of low or high voltage in circuits, and so on. The binary digits 0 and 1 are none of these things, in essence. Yet when manifested, binary digits have real effects. Programs are executed not only to calculate outcomes, but also to control steel mills and aircraft. Zeros and ones make things happen in the real world.

If that example strikes some people as artificial, there are also the genes that Brainard mentioned. The human genome may be written as a large "word" based on a four-letter alphabet—A, C, G, T. It may also be written as a huge number expressed in a four-digit notation: 0, 1, 2, 3. It occurs naturally as a sequence of amino-acid base pairs in the DNA molecule. This molecule breaks down when we die, and the expression disintegrates. Yet the number reappears in our descendants, expressed in new DNA. Where is the essential genome?

Essential content is ephemeral, appearing first in one manifestation, then in another. Although somewhat unreal in this sense, it is more than real in another, as though it had traded its cosmic reality for a new and more permanent mode of existence. Whatever its manifestation or expression, the same thing is always being manifested or expressed. As Pygonopolis pointed out in that delightful seafood restaurant in Izmir, essential content is also expressed by real objects and has very real effects in the so-called real world. If he decides to eat all the prawns on his plate, he will eat no more and no less than three. From this will flow many other consequences, including the time he rises from the table, his exact body weight, and a host of other, more subtle effects.

I think it is fair to say that what Pygonopolis meant by the holos is just the world of essential content, independent of the real world, including the world of particular mathematical expressions. Naturally, there can be no preferred mode for understanding the holos itself. The world of essential content is expressed equally well by "holos," "superior world," or "world of essential content." It is the invisible elephant.

Perhaps the most intriguing approach to the invisible elephant is the one described by Canzoni. When the cosmos is examined closely, matter turns out to be energy, and energy behaves according to mathematical dictates. If even the energy is unreal, only the structural information remains—the equations and formulas that describe everything. The cosmos vanishes, after a fashion. Was Pythagoras right? Is the cosmos made of number? One can scarcely entertain such an outlandish proposition.

More than any other name, I found *holos* best describes my own feeling about the invisible elephant. In contrast to the cosmos, where things have physical manifestations, the holos is where mathematics exists. What kind of place is the holos, though?

First, the holos is not necessarily in the cosmos. (I say "necessarily" only because I cannot rule out the possibility that Pythagoras was right and that we actually live in the holos.) Even if the holos is not in a place that we can physically identify, however, it has properties that qualify it for a kind of existence. Its landmarks of essential content, from numbers to theorems, persist like geographical features. Indeed, they have a permanent existence.

The holos may therefore be explored, as Pygonopolis pointed out, puzzles being the vehicle of choice for those with little or no mathematics. The mathematicians who have spent the past 3,000 or more years exploring the holos have amply demonstrated its independent existence, given the independent discovery, both in time and space, of a great many theorems. To speak of the reappearance of ibn Qurra's theorem about amicable numbers or the independent discovery of the calculus by Newton and Leibniz barely scratches the surface of this phenomenon. The essential content of mathematics is not created; it is discovered.

We are taught that to explain a phenomenon, we may propose any theory we like, but that simpler theories are preferred to complex ones. This is the principle of Occam's razor. There can be no question that the idea of a holos with a direct influence on the cosmos is an elaborate explanation, but, given the reality of mathematics in the cosmos, who can think of a simpler one?

The holos is home to the essential content of every number, every set, every string of symbols, every example of every kind of mathematical object—known, undiscovered, or undiscoverable. The holos houses the essential content of every theorem, every counterexample, and every mathematical statement, true or false. The place, however one conceives it, is immense. The total amount of information it contains is incomparably greater than the information that would seem to be required to specify the cosmos, even were the cosmos infinite.

Yet the holos subtly pervades the cosmos. What is an algebraic formula doing lurking in the wavelengths of the hydrogen atom? What gave Adams and Leverrier any right to expect that their predictions for the position of a new planet in the solar system would be correct? Who can doubt that if a cosmic system, be it planetary or atomic, obeys certain axioms, then it will obey every theorem that springs from those axioms?

Why, oh why, should the cosmos be structured like this? Perhaps there is no other way for a cosmos (qua cosmos) to be structured. Perhaps Pythagoras was right, after all.

Call me a fool, but never call me a coward. Having ventured this far from my comfortable world of accepted notions and tacit taboos, why should I not go all the way and venture an explanation for everything? The clues will come from an unlikely mix of Canzoni and Brainard.

Brainard, you will recall, believed that mathematics had an independent existence but only in a mind or, more properly, minds—and not necessarily human minds, either. Mathematics could not only be expressed by suitably programmed computers (every program being a kind of mathematical object), but it could also be discovered by computers, at least in principle. Taken in this general sense, we could say that mathematics exists independently in Mind with a capital "m," albeit not necessarily a conscious one.

It is tempting to imagine that the cosmos is like David Gridbourne's 2DWORLD program. Somewhere (not here), there is a great computer that runs the program 3DWORLD (or is it 4DWORLD?), and we are its denizens, trapped in a machine of some kind, a machine that we are unable to know the nature of—in principle! Such an explanation, however, smacks of that lazy tendency in us all to explain things by putting them off. For example, some would explain the origin of life on this planet by assuming that it drifted here from elsewhere in the form of "panspermia." Such an explanation merely puts the question off by forcing an answer from another quarter. In the present case, we must invoke something far more extraordinary than the holos—namely, a cosmic computer, and this I cannot do. My neck is already too far out.

Canzoni felt something was missing from physics, something hinted at by quantum mechanics. It had to do with consciousness. She finds the speculations of scientists such as Roger Penrose and Graham Cairns-Smith maddeningly vague, but holding great promise for the future. What if consciousness resides in a physical effect, as Cairns-Smith claims? In this view, the cosmos is literally permeated with consciousness, although it can only manifest itself in concentrated form where there is a brain or something equivalent (not necessarily a computer). The point is that such an all-pervading consciousness would presumably depend on matter or energy for its manifestation. Yet energy, according to Canzoni, is really information and therefore something of which the consciousness can be aware. This exploration is beginning to remind me of the ancient vedic symbol of the snake swallowing its own tail.

The cosmos exists because there is a mind that can think it. Does this mind also depend on the cosmos? Only the invisible elephant knows.

POSTSCRIPT

During the copyediting of this book, I received the sad news that Sir John Brainard had passed away gently one evening at his favorite pub, seated by the fire, puffing on his pipe and, I expect, pondering the connection between mathematics and mind. I wish him peace. Perhaps he finally has some answers.

I have also received word from al-Flayli, the Egyptian astronomer. He writes that in his view, Pygonopolis's holos and the Superior World of the Brethren of Purity are probably the same thing. His son Ahmed has just obtained a top scholarship at the Sorbonne. We will be hearing of him someday, no doubt.

During page proofs, a late flash had to be added. Maria Canzoni had gotten in touch with Pygonopolis, and he had flown to Venice to meet her. She writes, "At last I have a soulmate, one with whom I can share my theories. He has, so to speak, caught fire with these ideas, and we plan several joint publications. A thousand thank-yous for bringing us together!"

Something may yet come of the holos.

Index